THE SNOW GEESE

WILLIAM FIENNES

THE SNOW GEESE

PICADOR

First published 2002 by Picador
an imprint of Pan Macmillan Ltd
Pan Macmillan, 20 New Wharf Road, London NI 9RR
Basingstoke and Oxford
Associated companies throughout the world
www.panmacmillan.com

ISBN 0 330 37578 4 (HB)
ISBN 0 330 49285 3 (TPB)

1 3 5 7 9 8 6 4 2

A CIP catalogue record for this book is available from
the British Library.

Typeset by SetSystems Ltd, Saffron Walden, Essex
Printed and bound in Great Britain by
Mackays of Chatham plc, Chatham, Kent

for my mother and father

1: THE SNOW GOOSE

WE HAD NO IDEA the hotel would be the venue for a ladies' professional golf tournament. Each morning, before breakfast, competitors gathered at the practice tees to loosen up their swings. The women wore bright polo shirts, baggy tartan and gingham shorts, white socks, and neat cleated shoes that clacked on the paved walkways of the country club. Their hair was furled in chignons that poked through openings at the rear of baseball caps; their sleek, tanned calves resembled fresh tench attached to the backs of their shins. Caddies stood beside hefty leather golf bags at the edge of the teeing-ground, and the women drew clubs from the bags with the nonchalance of archers. Soon, rinsed golf balls were flying out from the tees, soaring high above the lollipop signs that marked each fifty yards down the fairway.

In addition to the golf course, heated swimming-pool and two tennis courts, hotel guests had at their disposal a peach-walled library, lit by standard lamps. White lace antimacassars lent fussy distinction to the dark red sofa and matching arm-chairs. Between the bookcases, in a simple giltwood frame, a colour print showed a suspension bridge, rigged like a harp, with staunch arched piers and high support towers lifting the curve of the main cables. Gold-tooled green and brown leatherbound books occupied the shelves alongside more modest clothbound volumes, the dye faded on their spines where light had reached it. The books were not for reading. Their purpose was to impart the atmosphere of an imperial-era country house. What the designer wished to say was, *This is a place to which gentlemen may retire with cigars.*

The shelves held arcane titles in strange conjunctions: an Anglo-Burmese dictionary next to a set of Sully's memoirs; G. Marañon's *La Evolución de la Sexualidad e los Estados Intersexuales* alongside Praeger's *Wagner as I Knew Him*; Carl Størmer's *De l'Espace à l'Atome* between J.R. Partington's *Higher Mathematics for Chemical Students* and the second volume of Charles Mills's *History of Chivalry*. An entire shelf was devoted to editions of the *Dublin Review* from the 1860s, containing such essays as 'Father de Hummelauer and the Hexateuch', 'Maritime Canals', 'The Benedictines in Western Australia' and 'Shakespeare as an Economist'.

One morning, after watching the golfers at the practice tees, I found a familiar book, a thin fawn volume almost invisible among the antique tomes. When I pulled *The Snow Goose* from the shelf, the books either side of it leaned together like hands in prayer. I settled back into an armchair and began to read, remembering how I had first heard this story, aged ten or eleven, in a classroom with high windows, sitting at an old-fashioned sloping desk with a groove along the top of the slope for pens and pencils to rest in, initials and odd glyphs gouged deep in the wood grain. Our teacher, Mr Faulkner, was a tall man with scant hair, flat red cheeks, and teeth pitched at eccentric angles. He wore silk paisley neckerchiefs and cardigans darned with wrong-coloured wools, and he kept his sunglasses on indoors for genuine optometric reasons. He was approaching retirement and liked to finish term with a story. One of the stories he read us was Paul Gallico's *The Snow Goose*.

I could feel, on the back of my head, the starched filigree imprint of the antimacassar. The library had no windows. Hotel staff wearing bold name badges walked briskly past the open door. I stopped noticing them. I imagined an Essex coastal marsh, an abandoned lighthouse at a rivermouth, and a dark-bearded hunchback named Rhayader, a painter of landscapes and wildlife, his arm 'thin and bent at the wrist like the claw of a bird'. Fifteen years had passed since I'd listened to Mr Faulkner reading this story, but its images rushed back to me: Rhayader's bird sanctuary; the October return of pink-footed and barnacle geese from their northern breeding grounds; Frith, the young girl, 'nervous and timid as a bird', who brings Rhayader an injured goose, white with black wing-tips – a snow goose, carried across the Atlantic by a storm as it flew south to escape the Arctic winter.

Rhayader tends to the snow goose. Time passes. The snow goose comes and goes with the pink-foots and barnacles. Frith gradually loses her fear of the hunchback; Rhayader falls in love with her but is too ashamed of his appearance to confess it. In 1940, startled by planes and explosions, the birds set off early on their migration north, but the snow goose stays behind at the lighthouse. Frith finds Rhayader loading supplies into his sixteen-foot sailing boat, preparing to join the fleet of civilian craft that would cross the Channel to rescue Allied troops from Dunkirk.

Much later, in a London pub, a soldier remembers details of that retreat: a white goose circling overhead as the troops waited on the sands; a small boat emerging from smoke, crewed by a hunchback with a crooked hand; the goose flying round and round above the boat while the hunchback lifted men from the beach and ferried them out to larger ships. The soldier compares the goose to an angel of mercy. He has no idea what became of the hunchback or the white bird, but a retired naval commander recalls a derelict small boat drifting between Dunkirk and La Panne, with a dead man lying inside it, machine-gunned, and a goose standing watch over the body. The boat had sunk, taking the man down with it.

Frith had been waiting for Rhayader at the lighthouse. He doesn't return. The snow goose flies back in from the sea, circles, gains height and disappears. A German pilot mistakes the lighthouse for a military objective, and blows Rhayader's store of paintings to oblivion.

I closed *The Snow Goose*, returned it to the shelf, and left the library for the fairways. But the tournament itself seemed lacklustre after the morning's pageant at the practice tees: the streaked blonde chignons; the easy rhythm of the swings; the fine baize finish of the green. Each caddy attended to his lady with devotion that verged on medieval courtliness: if she complained of a dirty clubface or perspiring hands, he would take one step forward, offering a fresh white towel. Sometimes women swung at the same time, and you could see two or three balls sailing out alongside each other, coterminous, holding still above the trees before inclining, as if by common assent, towards the flag.

*

I FELL ILL when I was twenty-five. I was a graduate student, working towards a doctorate. I went into hospital for an operation two days before Christmas. The surgeon did his rounds dressed as Father Christmas while a brass band toured the wards playing carols to requests. Between verses you could hear the bleeping of cardiac monitors and drip stands. I longed to go home. I heard a doctor tell another patient she could go home: he seemed to be granting her a state of grace. A few days later my mother and father picked me up and drove me home after dark, and I slept in a little room adjoining their bedroom, a room that my father had come to use as his dressing-room, but that had been my bedroom when I was very young. That night, I dreamed I was skiing. I was skiing on a wide open slope under blue sky, with no limit to the width or extent of the piste, and a sense of boundlessness, of absolute freedom. And then the snow had gone and a woman I had never met was leading me by the hand across a field, saying, 'Shall we go to Trieste? We must go to Trieste!' The window was ajar and a cold December draught blew through on to my head, and I woke up early, thinking my head was encased in ice. My mother swaddled my head in a folded blanket: I felt like an infant discovered in the wild and tended to by Eskimos.

I hoped that within two or three weeks I would be back at work, but there were complications. I went back to hospital for another ten days, and then my mother and father picked me up again and drove me home. I slept in the little room. I could smell my father's clothes. The bed was tiny: a child's bed. I slept on the diagonal, corner to corner, across the sag in the unsprung horsehair mattress, and when I woke up the first thing I saw was my great-grandmother's watercolour of Mount Everest, with a biplane flying towards the mountain, the word EVEREST embossed in black capitals on the cardboard mount. The picture hung above a table I'd always loved – it had a secret compartment, one flap hingeing where you least expected, with a knack to tricking the latch and always the same things inside: an old bible; pairs of cufflinks in a tissue nest; a clothes brush shaped like a cricket bat, the handle wound with waxy black twine.

There were further complications: hospital for the third time in as many months, a second operation. And then the need for serious convalescence – a few months, probably, for rest, for things to settle down, for my strength to come back. I gave up hope of meeting the university's requirements that year, and did not wish to be anywhere but home. My parents had moved into this house a few months before I was born: it had been the hub of my life, the fixed point. And now that everything had turned chaotic, turbulent and fearsome, now that I had felt the ground shifting beneath my feet and could no longer trust my own body to carry me blithely from one day to the next, there was at least this solace of the familiar. The house was my refuge, my safe place. The illness and its treatments were strange and unpredictable; home was everything I knew and understood.

A medieval ironstone house in the middle of England, miles from the nearest town. The stone was crumbling in places, blotched with lichens and amenable to different lights, ready with ferrous browns, ash greys and sunlit orange-yellows, with paler stone mullions in the windows, and a stone slate roof that dipped and swelled like a strip of water from gable to gable. A wood of chestnuts, sycamores and limes stood a stone's throw to the east, with a brook, the Sor Brook, running through the trees to a waterfall – a drop of nine or ten feet, with a sluice beside it, my handprints preserved in the concrete patch of a repair. Rooks had colonized the chestnuts, sycamores and limes, and when the trees were bare you could see the thatch bowls of their nests lodged in the forks, and black rook shapes perched in the heights, crowing like bassoons. The tall broach spire of a church poked the sky to the north, farmland drew away in a gentle upward grade to the south and west, and every one of these aspects – the wood, the farmland, the shape of the spire, the sounds of the rooks and brook – was a source of comfort to me. These things had not changed for as long as I could remember, and this steadfastness implied that the world could be relied upon.

I waited for my condition to improve. I wasn't patient. The edge of my fear rubbed off as the weeks passed, but I became depressed. In hospital I had longed to return to the environment I knew better than any other, because it was something of which I could be sure; because the familiar – the *known* – promised sanctuary from all that was confusing, alien and new. But after a while the complexion of the familiar began to change. The house, and the past it contained, seemed more prison than sanctuary. As I saw it, my friends were proceeding with their lives, their appetites and energies undimmed, while I was being held back against my will, penalized for an offence of which I was entirely ignorant. My initial relief that the crisis had passed turned slowly to anger, and my frustrations were mollified but not resolved by the kindness of those close to me, because no one, however loving, could give me the one thing I wanted above all else: my former self.

Leaves hid the nests in the tree crowns. Swallows returned in April, followed by swifts in May. After supper we'd sit out at the back of the house, watching swifts wheel overhead on their vespers flights, screaming parties racing in the half-light. Rooks flew in feeding sorties from the wood to the fields. You could hear the Sor Brook coursing over the waterfall in the trees, the sibilance of congregations saying *trespasses, trespasses, forgive us our trespasses*. But the sound was no longer a source of comfort. I couldn't relax into the necessity for this confinement. I felt the loss not just of my strength but of my capacity for joy. I tried to concentrate on the swifts, to pin my attention to something other than my own anxieties. I knew that generation after generation returned to the same favoured nesting sites, and that these were most likely the same birds we had watched the year before, descendants of swifts that had nested in the eaves of the house when my mother and father first moved to it; descendants, too, of swifts my father had watched as a boy, visiting his grandparents in the same house.

My mother suggested a change of scene, and we drove to a hotel close to the Welsh border. We had no idea it would be the venue for a ladies' professional golf tournament. Each morning, before breakfast, I walked down to the practice tees to watch the women loosen up their swings. I found *The Snow Goose* and read it straight through, remembering Mr Faulkner, the room's high windows, the grooved desks. I was suspicious of the story's sentimentality, its glaze of religious allegory, the easy portentousness of its abstract nouns, and I laughed at Gallico's attempts to render phonetically (as if they were birdsong) the East End speech of the soldiers in the pub and the upper-class diction of the officers. But something in the story haunted me.

*

MY FATHER LOVED BIRDS. A birdfeeder hung from a bracket at the back of the house: a long, thin tube of green wire mesh, chockfull with premium Hsuji peanuts. You could see the feeder through the French windows that opened on to the small paved terrace, and if you sat still you could watch coal tits, great tits, blue tits and sometimes nuthatches ransacking the store of red-husked nuts, the nuthatches easily picked out by their blue-grey backs and the way they clung (upside-down, tails-up) to the green mesh. I'd never paid them much attention as a child. A wren's habit of cocking its tail wasn't nearly as alluring as sport or pop music or television. I didn't want to listen when my father pointed out wagtails speedwalking across the lawn, chaffinches perched on the roof-angle, or the way a green woodpecker flew in bounds, folding its wings and losing height between bouts of flapping, so that you saw an undulation like someone stitching a hem and could say the name of the bird before you'd even made out the colour of its plumage.

But when we came back from the hotel, I wanted to learn about birds. I couldn't shake *The Snow Goose* from my head. I wandered round the garden, equipped with my father's Zeiss pocket binoculars and a simple beginner's field guide, looking for birds, trying to learn their names. Sometimes I'd describe a bird to my father and he'd name it for me: goldfinch, blackcap, yellowhammer. It must have been a surprise for my parents to see me showing these signs of enthusiasm: for months I'd been sullen, despondent, introverted, caught up in my own fears, resentful that my life had been interrupted so violently. In hospital, I'd longed to be at home. But by the end of May I was sick of it, restless, hungry for new experiences, different horizons. When I read Gallico's descriptions of the flights of geese, I wondered at the mysterious signals that told a bird it was time to move, time to fly.

I shared it, this urge to go. I was getting stronger. I was strong enough to be curious. It was as if I were trying to redeem my earlier failure to notice, the way I gave my attention, as I never had done as a child, to the swallows, swifts, rooks, wagtails, finches, warblers, thrushes and woodpeckers around the house – my father ready with a name, a habit, a piece of lore. I loved the swifts most of all. I'd never watched them so intently. My father said that after they left the house at the beginning of August many of them wouldn't land or touch down until they came back to nest the following May: they drank on the wing, fed on the wing, even slept on the wing. I thought of Gallico's snow goose flying south from the Arctic each autumn, the pink-footed and barnacle geese moving back and forth between Rhayader's sanctuary and their northern breeding grounds. Why did birds undertake such journeys? How did they know when to go, or where? How did swifts, year after year, find their way from Malawi to this house, my childhood home?

I was excited about something for the first time since I'd fallen ill, and I needed a project, a distraction, a means of escape. I carried books about bird migration up to a room at the top of the house, a real cubbyhole, tucked in under the roof, its low ceiling mottled with sooty drifts and rings, as if candles had smoked runes on to the cracked plaster – a room we knew as the eyrie, because it had the high, snug feeling of an eagle's nest. The pattern of fields I could see through the little two-light window was second nature to me, and I knew what each field was called: Lower Quarters, Danvers Meadow, Morby's Close, Allowance Ground. Sometimes swifts

screamed past the window as I sat in the eyrie, studying ornithology.

*

WE ARE TILTED. This was the first thing to understand. The
axis of the Earth's rotation is not perpendicular to the plane of
the Earth's orbit round the sun. It is tilted at about 23.5 degrees.
The tilt means that the northern and southern hemispheres are
angled towards the sun for part of the year, and away from the
sun for another part of the year. We have seasons. Climates turn
welcoming and inhospitable in regular sequence. Food supplies
dwindle in one place even as they burgeon in another. All
creatures must adapt to these cycles if they are to survive.
Migration is a way of coping with the tilt.

Hooded warblers, weighing a third of an ounce, fly more
than 600 miles non-stop across the Gulf of Mexico, and so do
ruby-throated hummingbirds, less than four inches long, their
wings beating twenty-five to fifty times a second. Red-footed
falcons fly from Siberia and eastern Europe, crossing the Black,
Caspian and Mediterranean Seas on their way to savannahs in
south-eastern Africa; demoiselle cranes fly over the Himalayas *en
route* to their Indian winter grounds; short-tailed shearwaters fly
from the Bering Sea to breeding colonies off southern Australia,
arriving each year within a week of the same date; the chunky,
short-legged waders called red knots fly all the way from Baffin
Island to Tierra del Fuego, an annual round trip of almost
20,000 miles. An Arctic tern, flying from the Arctic to Antarctica
and back again, might travel 25,000 miles in a year – a distance
roughly equivalent to the circumference of the Earth.

Six hundred thousand greater snow geese breed on north-
eastern Canadian Arctic islands and migrate south each autumn
over Quebec and New England to winter quarters along the
Atlantic from New Jersey to North Carolina. But these are far

outnumbered by the lesser snow goose, *Chen caerulescens caerulescens*, probably the most abundant goose in the world. The lesser snow occurs in two distinct colour phases. 'White-phase' snow geese have white plumage and black wing-tips; 'blue-phase' geese have feathers of various browns, greys and silvers mixed in with the whites, giving an overall impression of slaty, metallic blue. Blues and whites pair and breed together; they roost and migrate in mixed flocks. Both have orange-pink bills, narrower than the black bills of Canada geese, with tough, serrated edges for tearing the roots of marshland plants. A conspicuous lozenge-shaped black patch along each side of the bill gives them a grinning or leering expression.

Six million lesser snows breed right across the Arctic, from Wrangel Island off Siberia in the west, to Hudson Bay, Southampton Island and Baffin Island in the east, and at the end of summer they migrate to wintering grounds in the southern United States and northern Mexico. These are demanding, hazardous journeys of two or three or even four thousand miles, but the advantages of migration outweigh the risks. In the high Arctic latitudes, snow geese find large areas of suitable nesting habitat, relatively few predators, an abundance of food during the short, intense summers, and twenty-four hours of daylight in which to feed. And before the Arctic winter sets in, before their food supplies are frozen or buried deep under snow, they can fly south to exploit the resources and hospitable conditions of their winter grounds.

As I read, sitting in the eyrie, I kept thinking back to Gallico's story, Frith arriving at Rhayader's lighthouse with a wounded goose in her arms, either a greater snow goose or a white-phase lesser snow, knocked from its course by a storm as it flew south in its family group. I sought out photographs of snow geese: the wintry, laundered freshness of white plumage immediately after moult; the dense, lacquer-black eyes that glinted like china beads;

the wing bedlam of flocks rising from marshland roosts. I was drawn to these images. I felt shackled, cooped-up. It was as if I'd glimpsed birds through the high, barred window of a cell. Day by day, my restlessness intensified.

Then my father found an old map and left it on my bed in the dressing-room – a map of the Americas, rumpled and stained, worn through wherever foldlines intersected, with the flights of migrant birds streaking across it from one end of the continent to the other, Cape Horn to the Chukchi Sea. And the first thing to catch my eye was the long curve of watercolour green that represented the flight of midcontinent lesser snows, perhaps 5 million birds, from the Gulf coast of Texas north across the Great Plains towards Winnipeg; over boreal conifer forest and open tundra to Hudson Bay; and then on across the bay towards Southampton Island and a peninsula at the southern tip of Baffin Island called the Foxe Peninsula, or Foxe Land. I traced this route again and again across the map, dreaming of escape. Huge numbers of lesser snows nested in Foxe Land. One area, the Great Plains of the Koukdjuak, was said to support more than a million geese. What would they *sound* like, a million geese? What would it be like, I wondered, to see those flocks with my own eyes, coming into Foxe Land on the south winds?

I imagined a quest, a flight: a journey with snow geese to the Arctic. The pang of nostalgia, the intense longing to go home I had experienced in hospital, had now been supplanted by an equally intense longing for adventure, for strange horizons. I was as desperate to get away from home as I had been to return to it. I went back to the university at the end of the summer, but my heart was no longer in my work. I kept thinking of snow geese. I had been immersed in everything that was most familiar to me, that reeked most strongly of my past, and I was hungry for the new, for uncharted country. I wanted to celebrate my return from the state of being ill, find some way of putting the experiences of hospital behind me, the fear and shock of those weeks, the sense of imprisonment. I wanted to declare my freedom to move.

I booked a flight to Houston for the end of February, intending to find snow geese on the Texas prairies and follow them north with the spring.

*

THE DAY BEFORE I left for Texas, I took the train home from London. In the afternoon, my father and I went for a walk. A pink kite was snared in the churchyard yew tree; there were clumps of moss like berets on the corners of the headstones. We climbed a gate and strode out across Danvers Meadow, heading westwards, leaning into the slope, last year's sere beech leaves strewn through the grass. My father was wearing tan corduroy trousers and an old battered green waxed jacket; in one pocket he kept a matching green waxed hat in case of rain. We were walking at a steady pace, talking about the journey ahead of me, the rhythm of the walk going on under the words like a tempo.

A drystone wall ran along the ridge ahead of us, and we knew exactly what to expect from that vantage: gentle undulating country, a system of quickthorn hedges, stands of trees, fields ploughed or planted or left for grazing, and, beyond Lower Clover Ground, a cattle building with a corrugated roof, the herd's breath rolling out as vapour over wide steel gates. There were three straw bale ricks next to the building, with ladders and broken wood palettes propped against them, and further down the valley, beside the Sor Brook, stood a farmhouse with smoke rising from a brick chimney, a clutch of chicken sheds, a bunting of pink and white towels strung on a clothes line. This prospect was as familiar as our faces, as inevitable and apt, with spinneys, hedges, fields, slopes and the two buildings in their allotted places, each thing distinguished by a name: Hazelford, Buck

Park and Jester's Hill; Frederick's Plantation, Stafford Wood and Miller's Osiers; the Brake, the Shoulder of Mutton, the Great Ground.

We climbed a stile and walked on down towards the cattle building, the backs of my large black gumboots flopping against my calves. The drone of a twin-prop plane made us look up: a few cumulus clouds, purple-grey underneath, topsides gleaming like schooner sails; the furrowed white streamer of a contrail; the bounding flight of small birds. We heard the clang and judder of cattle on the steel gates, the herd breathing like organ bellows. A triangular sign said *Use Crawling Boards on this Roof,* and on the far side of the building there were grey feed troughs and wire fencing rolls, an open flatbed trailer, an old matt red Massey Ferguson combine and a heap of distressed farm machinery: ploughshares, harrows, iron scuffles, rusting discs and tines. Beyond the building the ground fell away to our left, down to the Sor Brook and the cricket-bat willows planted alongside it, their leaves a flashy bluish-green in summer. The brook ran past the farmhouse: a former mill, a tall, narrow building with white-framed windows under black timber lintels.

We passed the farmhouse, keeping to the high ground, with the Sor Brook meandering below us on our left side, and then we turned down the slope to the brook and walked back against the current through Keeper's Meadow and Little Quarters, the ground here disrupted by the red-brown earthworks of moles. Month-old lambs, and ewes with daubs of red paint on their haunches, grazed close to the quickthorn hedge; wool tufts were snagged on the quickthorn. There was a constant background chirrup and twitter, and at intervals the boom, quite far off, of a bird cannon. We walked side by side, opening gates and latching them shut, getting closer and closer to home. The spire came into view, the weathercock's tailplumes glinting in the low sun, and then the white stone chimneys of the house: our points of reference. There was no part of the world I knew so well, or loved so deeply. We walked up to the house, gravel crunching underfoot, taking our coats off as we approached the front door, the rooks garrulous in their high perches. I trod on the heels of the gumboots to get my feet out, and my father put one hand out against the wall to steady himself while he unlaced his boots.

Later, with the heavy red curtains drawn across the French windows, we leaned forward over my map of the Americas, following the flight of snow geese from Texas to Foxe Land. The mantel clock ticked; the fire snapped and puttered like a flag.

*

I FLEW TO HOUSTON, rented a metallic blue Chevrolet Cavalier and drove west towards Eagle Lake. It was exhilarating, just the thought that I should be in Texas, on my way to find snow geese, under my own steam, out in the world, in the new. And it *was* the new: coarse scrub prairie and fields ploughed for rice and sorghum running flat in all directions; windmills, cylindrical rice bins and galvanized farm sheds; pumping jacks going like metronomes in the blue half-light; and the silhouettes of mesquite trees like ancient Greek letters propped up on the narrow levees. I was already looking for geese: excitement – threshold alertness – warded off jetlag, and my eyes, like two small birds, flitted from place to place.

It was dark when I pulled up at the Sportsman's Motel, a single-storey, L-shaped building on the edge of Eagle Lake, pickups berthed at the room doors and a stuffed goose keeping vigil over the reception desk, wings akimbo. I hardly slept, my mind firing all night with anticipation, self-reliance, the prospect of Foxe Land.

A lady at the Chamber of Commerce had suggested I talk to Ken about snow geese. I called him from the motel that first morning.

'I'll meet you for lunch at the Sportsman's Restaurant,' he said. 'That's right across the lot from where you are. You can't miss it.'

Eagle Lake was a small prairie town with streets shaded by live oaks and magnolias, and white clapboard houses sitting in their plots like palaces in broad demesnes, boasting porches, decks, stoops, swingseats and awnings. American flags hung limply from peeling white poles. Red-winged blackbirds perched on the telegraph wires like notes on music staves. A railroad, the Southern Pacific, bisected the place, and every so often bells would ring out at all the level crossings, their clangour soon obliterated by the thunder of yellow locomotives and gravel-heaped gondolas plying the line from San Francisco to New Orleans, sending tremors rippling out under the buildings into the prairies.

I was the only person out walking. Hunters drove past me in pickup trucks, here to shoot ring-necked pheasants, sandhill cranes, chukars, mourning doves, scaled quail, white-tailed deer and the wild boar called javelinas, but especially the ducks and geese that wintered in their thousands on surrounding prairies. Waterfowl were spoilt for roosting ponds as well as rice and corn stubble in which to feed. The hunters, too, found everything they needed in Eagle Lake: Johnny's Sports Shop for camouflage gear, hunting knives, electronic bird calls, high velocity steel and premium tungsten *Tumble 'em with Tungsten!* shells, and Benelli, Browning, Remington, Winchester and Lakefield shotguns and rifles; guide services like Davis Waddell's Prairie Waterfowl Hunts or Lonnie Labay's Double 'L' Hunting Club; foul-smelling taxidermy studios to mount their canvasback, bufflehead or white-phase snow goose on a piece of Texas cedar driftwood and thereby make a trophy of it; and the Sportsman's Restaurant, across the lot from the motel, across Boothe Drive from the

Dairy Queen, to serve as their canteen, rendezvous and home-from-home.

Inside, the Sportsman's walls were hung with colour prints depicting hunting scenes: men setting decoys at dawn, sallies of startled ducks, the lope and fealty of labradors. Stuffed fowl took wing from driftwood mounts; miniature models of flying ducks dangled on fine chains from the hubs of ceiling fans: waitresses reached up and tugged on ducks to set the wooden blades turning. An angel, made of stained glass, hovered in one corner of the restaurant, the feathers in her wings suggested by intricate lead seams between almond-shaped pieces of smoky white glass. She wore a yellow dress sashed with a bolt of red cloth, and she carried a lamb.

The hunters wolfed down chicken fried steaks or chewed cuds of Red Man, Beech-Nut, Levi Garrett or Jackson's Apple Jack chewing tobacco, all kitted out in camouflage dungarees, shirts, jackets and caps, and necklaces of aluminium bands – identification bands removed from the shins of shot geese and strung on leather thongs. The foliage in their camouflage might be a crowded branching lit by little silver-grey leaves, or a pattern of palmate leaves like those of a maple or plane tossed in with the lobed leaves of oaks, or a design of fronds and the long, blade-shaped leaves of reeds and rushes, and sometimes men wearing these three distinct habitats sat together at the same table as if to illustrate the world's variety of bush and cover. The linoleum floor tiles were blotched with camouflage olive and khaki, and so were the tabletops, which meant that the hunters' arms disappeared when they reached for the salt cellar or sugar jar, the patterns and colours of their sleeves getting lost in those of the tables.

In flat-toned, drawling voices they exchanged stories of endurance and derring-do.

'I was hunting one time in Colorado. Three days we rode up and down those slickass mountains, looking for elk. Didn't see squiddly. Driving home, killed a deer with the truck. Caught it square on the fender. Knocked it dead.'

'Least you got something.'

'I used to deep-sea fish whenever I was able. I never got sick. I got queasy once, but I never got sick.'

'Oh, I got sick. I got real sick.'

'Only time I got sick was fishing with my doctor. I had to look at him eating pickles. That made me nauseous, looking at the guy eating pickles. Doctor gave me some tablets. Said, "Here, take these." So I took the tablets and – *boom* – slept for six hours, right there on the deck of the boat, face upward. You know how hot the sun gets on the Gulf? When I woke up I was chargrilled. Blistered like a fish. You've never seen blisters like these. Oh boy!'

'One time I took these guys ice-fishing in Iowa. Three mafia bosses. Sunglasses, camel-hair overcoats, cigars. Drove out in a black limousine with smoked windows and a chauffeur. Chauffeur drove this limousine right over the lake, pulled up where I'd drilled through with the auger. Bosses had him wait there in the limousine while they jigged for bluegill!'

Ken saw me standing at the door, scanning the tables, and beckoned me to come over and join him. He was sitting alone, wearing a camouflage jacket and a baseball cap that declared his allegiance to the Dallas Cowboys – a short, compact man of about forty, stroking his ginger goatee beard between thumb and forefinger as though thinking deeply on some chain of cause and effect. His contact lenses were tinted electric sapphire blue; his pupils were set like peppercorns in rings of blue; these strange eyes glowed with significant candlepower. A stuffed white-phase snow goose was mounted on the wall behind him, wings spread wide. The black tips to the wings weren't decorative: the concentration of melanin pigments – the pigments responsible for dark colouring – strengthens the primary flight feathers, making them more resilient, an adaptation often seen in birds that undertake long migrations.

'Glad to see you,' Ken said. 'Take a load off and have a seat.'

He had a small rice farm attached to his grandfather's ranch; as a sideline he acted as a guide for hunters. We ordered lunch and talked about snow geese. The numbers of lesser snows had been causing concern, Ken said. The population was growing beyond all predictions. Two thousand pairs nested at the Cape Henrietta Maria colony in 1960; 225,000 pairs nested there in 1998. A hundred years ago, lesser snow geese wintering in Texas and Louisiana kept to coastal marshes between the Mississippi Delta and Corpus Christi, Texas, feeding on the roots and rhizomes of rushes and marsh grasses. But with the agricultural development of the Gulf and Great Plains states, large numbers of geese had begun to winter further north on inland prairie habitats, incorporating waste rice, corn and barley grains into their diet. The geese were now finding an almost unlimited supply of food between Winnipeg and the Gulf of Mexico. This abundance, together with the establishment of wildlife refuges where geese were safe from hunters, had resulted in a population explosion, with so many snow geese reaching Arctic breeding grounds that tundra habitats were being stripped bare, the vegetation unable to recover from year to year. Along 1,200 miles of coastline surrounding Hudson Bay and James Bay, scientists estimated that more than a third of the original tundra habitat had been destroyed, with another third seriously damaged.

'We call them wavies, these snow geese,' said Ken, adjusting his cap, 'because, when they fly, they kind of wave up and down. I expect those birds in October, maybe November. They find a pond to roost on, a few thousand to a roost, and take off at

sun-up for breakfast, you can set your clock by it. They leave
the roost for the fields to have themselves a feed, and during the
middle of the day they dawdle and rest up, and in the afternoon
they get busy again, and then they go back to the roost at sun-
down. It's like clockwork. In and out. So if you're hunting geese,
you'll get out there when it's dark in the morning, put out
decoys and wait for them to fly over. You can call those birds in
with a voice call or an electronic call, whichever you prefer. And
normally they'll be off again at the beginning of March, but this
year a lot of geese have gone already because it's so warm, we've
got an early spring. We've got a few geese left on the land but
those birds are laggards; those birds, shall we say, are bringing
up the rear.'

Ken suggested I meet him again at the Sportsman's later that
afternoon. He said I could follow him out to the farm. He'd
show me the roosting pond; I could wait for sunset and watch
the birds fly in. I was impatient for my first sight of snow geese.

We met at five o'clock. Ken was stroking his beard between
thumb and forefinger.

'You ready to see some birds?' he asked.

His tan Dodge pickup had dents and blemishes all over its
body, like a rally car. Driving the blue Cavalier, I followed
the Dodge out of the Sportsman's parking lot on to Boothe
Drive. We drove slowly out of Eagle Lake, into the prairie, the
sky arching from horizon to horizon, a vast inverted bowl of
glass, clear blue but for two parallel bars of feathery cirrus. We
turned off the blacktop on to a dirt farm track, dust billowing
from the pickup's back wheels. I kept looking to left and right,
eager to see snow geese. The only vertical accents were the
telegraph poles, rice bins and radio masts. The square fields were
ploughed or left as pasture. The tracks made a grid between the
fields and we worked our way along them in knight's moves
until we came to a small two-storey house, a box set down in the

22

middle of the prairie, with a staircase running up the outside to a door on the upper floor. We parked our cars at the foot of the stairway.

'My grandfather built this house for himself,' Ken said. 'It's like a retirement place. He builds a house, and then he decides he wants to live by a lake. So he builds a lake right next to it. Raised dykes, pumped water into it, stocked it with bass for fishing – an actual lake, above ground level, right on top of the prairie. Come inside.'

We walked up the stairway. Inside the front door was a welcome mat that said, *Wow! Nice underwear!* Coats hung from a rack of antlers. The living-room had an open-plan kitchenette at the back, and sliding glass doors that opened on to a wood deck with a view of the raised lake and surrounding rice fields, one of which had been flooded to make a roost for geese. The water was only a foot or two deep: a few ducks were floating on it; waders picked their way along the muddy edges. Ken slid back the glass doors and we walked out on to the deck. The sun was low; the air was cooling quickly. Inky Black Angus cattle were lounging in scrub grazing fields and shambling along levees. Ken's electric blue eyes glowed more intensely in the dusk light. There was no wind to spin the tricolour slats of the windmills.

Jack, Ken's grandfather, had tailored the house to his exact specifications. He wanted a deck because he liked to sit and watch geese and ducks fly in to their roost at sunset. A boardwalk ran from the deck to the edge of the lake, with a wooden raft floating on pontoons at the end of it, furnished with chairs for fishermen. The house was dedicated to birds. Door handles were moulded and carved to resemble herons' beaks; a miniature ceramic Canada goose glided on a knot at the bottom of the bathroom light-pull; and all over the walls were prints and watercolours of waterfowl and hunting scenes, along with photographs of hunters tricked out like militiamen, camouflaged from head to toe, toting rifles, holding out braces of dead geese, red daubs stark on the bodies of white-phase snows. Here were old wooden duck decoys displayed on varnished shelves; glass ashtrays engraved with Canada and snow geese; books on decoys, waterfowling and the sporting life standing upright between sturdy bookends, each bookend the head, neck and shoulders of a brass duck. A laminated map of the world was spread out on the floor, with stacks of glossy hardcover bird books placed at each corner to stop it curling or rolling up: the world kept in shape by the weight of all the birds in it.

'Here's Jack,' said Ken. We watched a pickup tear between two fields, dragging a dust cloud. It pulled up beside the house, we heard footsteps on the stairs, a short man with trim white hair and stocky, bundled vigour muscled into the room, hung his bomber jacket on the antlers and strolled out on to the deck. Jack's skin was deep brown, with a rough pimpling on his neck as though feathers had been plucked from it. He'd been repairing fences: dirt was grained in the creases of his hands; his palms showed maps of river deltas when he opened them in gesture. Before becoming a farmer he'd been a pilot in the US Air Force, and he pointed proudly to photographs of his younger self in uniform, sunlit, beside fighter planes – photographs that didn't seem out of place among the pictures of birds, because all had to do with flight, skies, the genius of wings.

'This guy's come from England to watch geese,' Ken said.

'Is that so?' Jack replied absently, smoothing his hair back, gazing out over the lake and flat fields.

'He's going to follow them from Eagle Lake to Canada, Hudson Bay, maybe even the Arctic Ocean.'

'Each to his own,' said Jack.

'He just flew in. Hasn't ever seen a snow goose.'

'Is that right? Sometimes I wish I'd never seen a snow goose.'

'Why do you say that?' I asked.

'Too many of the damn things. Stick here a while, you'll see more of those birds than you can shake a stick at. May not look like it right now. Wait for that sun to go, all hell breaks loose.'

Jack didn't stay long. He retrieved his jacket from the antlers and paused at the door.

'Sounds like a wild goose chase,' he said with a smile, eyes twinkling. 'OK, Kenny. I'm gone.' He went.

On the deck, in the fading light, Ken pointed to a track that ran along one side of the flooded field. I was still scanning the sky, checking the horizon, willing birds to appear.

'All you have to do is park there and sit tight,' Ken said. 'The geese'll fly right in on top of you. Just sit tight. Be patient.'

I followed him back through the living-room and down the outdoor staircase. Ken locked up the house. We shook hands at the bottom of the stairs.

'I'll see you in town,' he said.

Ken drove off in the Dodge, leaving me alone on the prairie. It was just after six o'clock. I parked the Cavalier at the edge of the flooded field and waited, tense, eyes keen, vigilant for geese. I lifted my binoculars and panned across the water, finding ducks floating in twos and threes, waders tottering as if on stilts at the edge of the pond. In front of the sun the birds were silhouettes, and I was too much of an amateur to tell one species from another. But when I saw eight tall, slender birds with the long necks, legs and bills of herons, and shaggy tail bustles, and the dainty gait of ballerinas, I knew instantly that they were sandhill cranes, the oldest species of bird in existence, known to have lived in Nebraska in the Miocene, 9 million years ago – birds which, it was once believed, helped smaller birds migrate by carrying them on their backs. These sandhill cranes would themselves soon be leaving for Arctic Canada, staging in Nebraska's Platte River on their way to breeding grounds between Alaska and Hudson Bay.

The sun was close to the horizon now, not the source of
light but the point to which all light was gathered, as if the day
were going home. I leaned back against the car, on the brink of
geese, my ears tuned, my eyes alive to the slightest movement.
Ducks muttered on the shallow water. Red lights glimmered like
cigarette tips on the radio masts. The mesquite trees had the
bare, stony branchings of tree corals. I heard bells pinging in
Eagle Lake, several miles to the north-east, and then the rumble
of a freight train, the ground vibrating with its industrial
repercussions. There was a pale streaked redness in the west, but
the rest of the sky was a deep liquid Prussian blue, with a pair of
bright stars appearing very close together in the south-west:
Venus and Jupiter in conjunction.

A bird approached the pond from the south-west: a heron, a
great blue, easy to distinguish from a crane because herons fly
with a pleat in their necks, heads retracted on to shoulders, while
cranes stretch their necks out straight, without a kink. Sometimes
we came across solitary grey herons standing like baptists on the
banks of the Sor Brook or at the edge of a pond, footed to their
own reflections, and my mother had painted one – its yellow,
scabbard-shaped bill and eye, the wispy black plumes at the back
of its head – on a strip of old rollerblind that hung in the
bathroom, the window looking out at trees with rooks cawing
hoarsely in their heights, a kingfisher of smoky, chipped glass
standing on the sill beside a tin tray of quartz, pumice and agate
pieces, the white wall to the left of the door marked with initials,
dates and horizontal dashes: children's heights, measured year by
year, heels to the skirting board. This great blue flew right over
the holding pond, a ray ghosting through sea water, with five
American white pelicans following behind, heads retracted like
the heron's, gular pouches sagging like jowls under their long
bills. It was half past six. I leaned back against the blue car,
waiting.

The first sign was a faint tinkling in the distance, from no particular direction, the sound of a marina, of halliards flicking on metal masts. Drifts of specks appeared above the horizon ring. Each speck became a goose. Flocks were converging on the pond from every compass point, a diaspora in reverse, snow geese flying in loose Vs and Ws and long skeins that wavered like seaweed strands, each bird intent on the roost at the centre of the horizon's circumference. Lines of geese broke up and then recombined in freehand ideograms: kites, chevrons, harpoons. I didn't move. I just kept watching the geese, the halliard yammer growing louder and louder, until suddenly flocks were flying overhead, low over the shoulder, the snow geese yapping like small dogs, crews of terriers or dachshunds – urgent sharp yaps in the thrum and riffle of beating wings and the pitter-patter of goose droppings pelting down around me. They approached the roost on shallow glides, arching their wings and holding them steady, or flew until they were right above the pond and then tumbled straight down on the perpendicular. Sometimes whole flocks circled over the roost, thousands of geese swirling round and round, as if the pond were the mouth of a drain and these geese the whirlpool turning above it. Nothing had prepared me for the sound, this dense, boisterous din, the clamour of a playground at breaktime, a drone-thickness flecked with high-pitched yells, squeals, hollers and yawps – the entire prairie's quota of noise concentrated in Jack's holding pond by the two-storey house and the raised lake stocked with bass for fishing. I breathed it in. It was seven o'clock. There was a half moon. I waited until the birds were settled, then drove back slowly along the farm tracks, leaving the headlights off until I reached the highway.

2: AUSTIN

FOUR EVENINGS IN A ROW I waited for geese at Jack's roosting pond. I sat on the bonnet of the Cavalier, flicking through field guides or looking north, imagining the Great Plains stretching away to Canada, and all the snow geese already on the move, bound for traditional staging areas in Nebraska and the Dakotas. There were always a few ducks on the water, a few sandhill cranes foraging with lanky grace in the ploughed land at the edge of the pond, and herons and pelicans, and shambling longhorn cattle, and red lights glowing in strict, linear constellations on the radio masts. My pulse quickened as the same thousands of geese converged on the roost. I took shelter inside the car, wise to the turd squalls.

The snow geese had been gleaning for leftover grain and grubbing for the roots of sedges and grasses. They were about to depart on a migratory journey of around 3,000 miles: they had to have sufficient energy reserves for the flight ahead. Twice a year, before they migrate, birds go through a period of intensive eating, or hyperphagia, during which they lay down deposits of subcutaneous fat: essential fuel, providing twice as much energy per unit of weight as carbohydrate or protein. Some birds, like the blackpoll warblers which leave the coasts of Nova Scotia and New England on 2,000-mile non-stop flights over the Atlantic to the north-east coast of South America, almost double their weight before departing on migration.

Many birds winter close to the equator, where there are no reliable seasonal variations (shifts in day-length, temperature or the availability of resources) to provoke such changes in behaviour. Yet these birds fatten, and depart for their breeding grounds, at appropriate times. When Eberhard Gwinner kept willow warblers in temperature-controlled chambers with constant cycles of twelve-hour days and twelve-hour nights from the end of September, they still moulted and came into migratory condition in the spring at the same time as control birds kept in cages on their African winter grounds.

The warblers were not relying on environmental cues: the changes in their behaviour were prompted by an internal clock consisting of two fundamental rhythms. Circadian rhythms, corresponding to the twenty-four-hour cycle of the Earth's rotation on its axis, regulate daily changes in metabolic rate, body temperature and level of alertness. Circannual rhythms, corresponding to the annual cycle of the Earth's orbit round the sun, control changes in behaviour associated with reproduction, moult and migration. These rhythms are not exact, but are tied to the natural day and year by external indicators called Zeitgebers, from the German for 'time-givers'. The most important Zeitgeber is photoperiod: the amount of daylight in a given day.

A snow goose, like all migratory birds, inherits a calendar, an endogenous programme for fattening, departure, breeding and moult. This schedule is essentially fixed, but it can be fine-tuned by environmental conditions. Due to the early spring, snow geese had been leaving the prairies around Eagle Lake a week or two sooner than Jack and Ken had come to expect. Millions of geese had already left their winter quarters for Arctic breeding grounds, and the flocks roosting by Jack's house would soon be following them. Each evening, I'd driven back from the holding pond to the motel elated, songs blasting, the wild lung-top gabble of the flock still ringing in my ears. And I became restless too, eager to be on the move, to be covering ground, working north towards Foxe Land.

I found Ken in the Sportsman's Restaurant, stroking his

ginger beard like a sage. I thanked him for his help; he wished me luck.

Outside, heat haze rose off the asphalt like a version of water.

*

ELEANOR WAS SIXTY-SEVEN, a small woman with a bird's light bones. Her soft white hair had the airy, fluffed-out quality of down feathers, and she raised her hands occasionally to pat at it with open palms, attending to the outline. Crow's feet rayed out from the corners of her eyes, deepening when she smiled, and there was a red flourish in both her cheeks like the smudges left on bats by new cricket balls. When she greeted me at the door of her house she was wearing pastel-blue cotton trousers with an elastic waistband, and an old green sweatshirt that served as the plain setting for an elaborate brooch: a tin plate equipped with knife, fork and spoon.

'Welcome!' she said.

We hadn't met before. Her nephew, whom I'd met in England, had put us in touch, and Eleanor had offered me a place to stay if I passed through Austin on my way north with snow geese. She lived in a residential district west of the centre of Austin, in a single-storey house of pink-tinged Arkansas ledge stone, with a neat shallow-gradient slate-tile roof and a basketball hoop still screwed into the wall, though her son had long since left home. She had been married to an architect; they had designed the house together. Nearby houses strained for peculiar grandeurs: mock, ivy-covered castles with turrets and arrow-slit windows; plantation-style villas with white Palladian porticoes; soft-cornered adobe bungalows complete with protruding pine roof beams and ristras of red jalapeño peppers. Young mothers pushed children in streamlined prams under the evergreen live oaks. Driving with the window down, I could hear the sibilance of lawn sprinklers and the harsh, cracked-whistle calls of grackles, and when the sprinklers came into view there were slips of rainbow caught in their ambits as if in the finest fishing nets.

'Come inside,' Eleanor said. 'We'll get you settled in.'

The walls were panelled with walnut: the living-room had the mild light and coolness of a glade. Dishevelled oriental rugs lay on polished bare wood floors. Sunlight, filtered by trees, entered through sliding glass doors. There was a sofa upholstered in faded mulberry cloth, and a well-worn leather armchair bearing Eleanor's precise indentation, with an Anglepoise lamp on an end-table beside it. Black-and-white photographs hung in silver frames: tuxedos, evening gowns, brides and grooms, diploma scrolls. A black upright piano stood against the far wall, a volume of Bach open at a polonaise, and opposite the piano stood a threefold Chinese screen with trees, foliage and exotic birds in enamel and mother-of-pearl on a black background, and a column of Chinese calligraphy on each of the three panels.

'It's from a dynasty,' Eleanor said. 'Though search me which.'

Between the screen of birds and the sliding windows was a hip-height round wooden table covered with tortoises. Eleanor collected them on her travels, and she'd placed them carefully on the table, evenly spaced, all facing in the same direction, towards the piano at the far end of the room. Some had tails and some didn't. There were delicate glazed grey, green and blue ceramic tortoises; chunkier, rough-clay tortoises; invertebrate fabric tortoises with beanbag fillings; dirty brass or steel tortoises, their shells inverted to make ashtray bowls; leather tortoises conceived as purses, with zips along the back or side; tiny glass tortoises, like raindrops with limbs; and tortoises carved from hard, dark woods, with grid patterns seared into their shells by red-hot pointed tools. One of the tortoises had a smaller, baby tortoise crawling up over its shell, and one was a jelly mould made of lightweight metal.

'My mother gave me my first one,' Eleanor said, taking a

medium-sized grey ceramic tortoise from the middle of the parade. 'After that I looked out for tortoises. Antique stores, markets, bazaars. If I go away, I always try to find a tortoise.'

To enter the walnut-panelled corridor that led from the living-room to the bedrooms, you had to pass through swinging walnut doors, slatted like Old West saloon doors.

'This is you,' Eleanor said, showing me into the small guest bedroom. 'I'll leave you for a minute.'

A birdcage on the chest-of-drawers: an elegant antique birdcage of thick brass wires that curved together at the top, forming a dome, with a rod for perching and a brass ring for hanging the cage at the zenith, and everything in the room subordinate to the cage, the clean, bright wires enclosing the not-here of the bird inside it. I put my bags down and contemplated the room: the same glade atmosphere, with wood on the walls and underfoot, the floorboards part-covered by a needlepoint rug in bold dahlia colours, Eleanor's handiwork. I unzipped my bag, took out my bird books and binoculars, and put them on the chest-of-drawers, next to the empty birdcage.

I took it all in, this new place: its colours and textures, its different lights, its things. Above the bed hung a drawing of a cowboy, riding a horse at a rodeo. The horse was executing a buck, poised on its hind legs, head and neck angling downwards, front legs about to hammer the ground. The cowboy was airborne, bounced a foot from the saddle. One hand held the reins; the other was held out over the horse's head. You could see that when those front legs hit the ground, the cowboy would slam back into the saddle and lose his balance, jolted, but for now all was poised and beautiful – the rider's flaring leather chaps, adorned with rosettes; his saddle's horn, skirt, cantle and bucking rolls; the tapaderos on the stirrups; the leather fenders. A small patch of shadow on the rider's Stetson indicated the depression where his fingers would grip to lift the hat to a lady.

'Would you like some tea?' Eleanor asked, peering into the room.

We passed back through the swinging saloon doors like gunslingers. I followed Eleanor into the kitchen, and watched as she put some water on to boil, a spiral glowing orange on the electric hob. She didn't have a kettle. She had two cast-iron skillets, one six inches across and one nine and a half inches across, and used these to boil water for tea, the smaller skillet holding just enough water for a single mug, the larger just enough for two. She handled the skillets using one of two quilted oven pads, a black sheep and a white sheep, their fleeces browned and singed through by burns. Her birdlike lightness and the rugged iron heft of the pans seemed imagined by two radically different minds.

'I'm so glad you're doing this,' she said, watching the water in the skillet. 'I love being a part of this adventure.'

'Good!'

'I just read *The Snow Goose*. I wanted to have an idea of what set you off on this I don't know what you'd call it. And it's so sad!'

She was holding the skillet's handle with the black sheep, watching the water.

'Sometimes I think it's amazing that water boils,' she said. 'All those little bubbles suddenly appear. What tea would you like?'

She opened a cupboard to the right of the cooker: it was crammed with boxes of loose leaf teas and tea bags – black teas, green teas, exotic herbal infusions and improving, medicinal tisanes, their bouquet wafting from the open cupboard.

'There's no shortage of teas,' she said.

We both chose peppermint. Eleanor returned the black sheep to the cooker rail: the two sheep flopped over the rail like people touching their toes. Magnets held postcards, photographs and cartoons to the white fridge: pictures of pianos, virtuoso pianists, piano lessons, sheet music. The magnets themselves were treble clefs, bass clefs, quaver pairs and miniature grand pianos. In the middle of the fridge, held to the metal by two pianos, was a piece of card on which Eleanor had copied out a proverb: *It takes both rain and shine to make a rainbow.* Her handwriting took a vine's delight in winding and spiralling tendrils.

I followed Eleanor out through the sliding glass doors on to a roofed balcony that jutted into crowns of elms, live oaks and pecans. A set of wind chimes hung in one corner – six metal tubes of different lengths, like tubular bells, with a wooden puck in the middle, attached to a square metal sail. The sail got wind of the faintest breeze and carried the puck from bell to bell, sounding low, hollow notes. Squirrels ran along the boughs; glossy, purplish-black grackles bungled noisily in the leaves.

We sat down at a glass-topped table and jigged our teabags by the strings.

'So how about these geese?' Eleanor asked.

I told her about the roost at Eagle Lake, the sunset returns of blue-phase and white-phase snow geese. Millions of birds, I said, were already coursing up the flyways, across the Great Plains, towards Manitoba. They were heading for traditional staging areas in Nebraska's Platte River valley, the lakes of South Dakota and North Dakota, and the grainfields west of Winnipeg. After resting, and replenishing their fat stores, they would push north with the leading edge of spring towards nesting grounds along the edge of Hudson Bay, and, further north and east, on Southampton Island and Baffin Island. I mentioned the tilt of the Earth, circannual rhythms, the period of intensive feeding that precedes migration, the twice-yearly restlessness that prompts birds to undertake such ambitious flights.

In accordance with their inherited calendars, birds get an urge to move. When migratory birds are held in captivity, they hop about, flutter their wings and flit from perch to perch just as birds of the same species are migrating in the wild. The caged birds 'know' they should be travelling too. This migratory restlessness, or Zugunruhe, was first described by Johann Andreas Naumann, who studied golden orioles and pied flycatchers at the beginning of the nineteenth century. Naumann interpreted Zugunruhe to be an expression of the migratory instinct in birds.

Circannual rhythms control the onset of Zugunruhe; restlessness prompts the birds to depart. Migrants do not need to rely on the example of parents or other experienced individuals. In some cases, birds do not even have such examples to follow. Eurasian cuckoos lay their eggs in the nests of other birds, obliging surrogate parents to rear their chicks. As soon as they have laid their eggs, adult cuckoos are free to migrate: most fly south in July. Young cuckoos migrate about a month after their true parents. Rather than following their foster parents, these juveniles fly by their own instincts to join adult cuckoos in African and south-east Asian winter grounds.

Both the need to migrate and the route to be taken are at least partially coded in the genes. Caged birds do not only get restless, they get restless in a particular direction. Each autumn, populations of garden warblers migrate from Germany to Africa via Spain and Morocco. They fly south-west to the Strait of Gibraltar, then head south and south-east to their winter grounds. Garden warblers raised in cages under constant environmental conditions begin to exhibit Zugunruhe just as their counterparts in the wild are setting off on migration. The caged warblers show distinct directional tendencies. They hop towards the south-west while free warblers are flying south-west across France and Spain. The caged birds then change direction and hop south and south-east just as the free birds are turning south and south-east over Gibraltar. They seem to act out the entire course of the migration in the confinement of their cages.

Birds, it seems, are genetically programmed to fly in a certain direction for a certain period of time. If these inherited instructions are carried out, the bird will arrive in its breeding or winter grounds. But migrating birds must have a way to compensate for unpredictable environmental conditions: individuals flying with the wind behind them cover much greater distances than conspecifics flying into headwinds; storms may blow them off course.

The flexibility of a bird's innate programme was demonstrated in the late 1950s by the Dutch ornithologist A.C. Perdeck, who captured more than 11,000 starlings near The Hague during their autumn migration from breeding grounds in north-west Europe, following the North Sea coast in a general south-west direction towards winter quarters in Holland, Belgium, north-west France, Ireland and southern England. He ringed every bird, having first identified its age and sex.

The starlings, loaded in bamboo cages, were immediately flown to Switzerland and released near Basel, Zürich and Geneva. Three hundred and fifty-four were recovered. Perdeck found that most adult birds had flown from Switzerland on north-westerly courses, and that some had actually reached their normal winter grounds. But the juveniles tended to fly south-west, ending up in southern France, or even Spain. Inexperienced starlings were not able to adjust to the displacement. In adult birds, the endogenous migratory programme was modified by experience, while juveniles flew on a fixed heading, with no regard to circumstance.

'My husband watched birds,' Eleanor said. 'I can tell you such and such is a grackle and such and such a hummingbird. I put out a feeder for the hummingbirds. But I wouldn't call myself a birdwatcher in so many words.'

'I wouldn't call myself one either, not in so many words.'

I told Eleanor about my illness, my long convalescence at

home; how, after finding *The Snow Goose*, I'd paid attention to the birds around the house for the first time, asking my father their names.

A breeze lifted; the wooden puck swung from chime to chime. We could hear the traffic on Lamar Boulevard.

'I don't mind the traffic,' Eleanor said. 'Sometimes you hear the car wrecks, which is the only thing I don't like about this balcony. I sit out here all the time. It feels like a nest. You're right in the trees.'

The doorbell rang.

'That'll be my son, Matthew,' Eleanor said. She got up, slid back the glass doors and slipped through into the living-room, returning with a man in his late thirties, much taller than his mother, dressed in shorts, a sun-faded red baseball cap and blue T-shirt, with a strong, tanned, outdoorsman's face, brown eyes, and a pronounced brow like a girder running from temple to temple.

'How are the geese?' he asked me.

'Fine, I think. They're starting to move.'

'I've been seeing them fly over the house. Sometimes I don't see them, I *hear* them.'

'Matthew's building a house,' Eleanor said.

'Really?'

'Yeah. I'm doing a lot of the work myself. It's pretty hard. I've got a couple of guys helping me. Right now, I don't like to be away from the site for too long. I like to be around to keep an eye on things. I'm the only person who knows what this house is actually going to look like. It's all in *here*.'

He tapped his temple with his index finger.

'My husband and I built this house together,' Eleanor said. 'It's quite an undertaking.'

'It's pretty hard,' Matthew said again. 'It's exciting. I've lived all over the place. I got tired of moving around. I moved twenty times in the last eight years. I just thought, "Enough's enough." A piece of land came up for sale and I snapped it up. It's in the hills, just outside the city, right under this radio mast. It was cheap at the price. It's the radio mast: you can hear it zinging. People thought they'd get their heads fried.'

'You don't notice it for long,' Eleanor said.

We heard a siren approach and fade on Lamar Boulevard.

'You should come up and see the house,' Matthew suggested. 'There's really something there now that you can say, "That's a house."'

We agreed to visit his house the next morning.

*

MATTHEW DROVE A Jeep Cherokee; his mother drove an old caramel Mercedes with slippery leather seats and a sticker in the back window saying, *There's No Place Like Narnia*. The road rose and wound through forests of cedars and small oak trees. We caught glimpses of a radio mast, footed in evergreens.

'You can't get lost,' Eleanor declared. 'You just head for the mast.'

Matthew's drive was marked by a mailbox.

'He had an address before he had a house,' Eleanor said.

We turned off the road on to a rutted dirt track that led through cedars to a clearing: an oval of rough ground, razed by bulldozers, with earth, rubble, stumps and construction debris driven up in heaps and ridges, and a charge of fine dust in the air, tasting dry on the tongue. The sun was hot; there were few clouds. The bone structure of a neat two-storey house stood on a concrete foundation slab at the centre of the clearing – a frame of steel I-beams half-clad in grey concrete sections. No roof. Spaces left for windows and doors. The ends of threaded steel rods poked from the concrete, and timber gables were stacked on the rough ground at the near edge of the foundations, waiting to be hoisted to the top of the frame. It was the idea of a house, not quite transfigured into the thing itself.

Matthew was standing by the stacked gables, a utility belt slung round his hips, laden with builder's gear: hammer, measuring tape, pencils, a medley of nails and screws. Beyond the house, the mast drew your gaze upward like a spire. It was painted red and white, and stayed by taut steel hawsers. Two turkey vultures, wing-tip feathers spread like fingers, glided in wide, slow circles over the cedars, close to the hawsers. Eleanor and I walked over the rough ground towards the house, and as we approached, Matthew raised his right arm and pointed upward, as if he were another, smaller mast.

'Check it out,' he said casually.

'What?' Eleanor asked.

'Take a look.'

We all looked upwards. I didn't see them immediately. The snow geese were flying high, each bird catching the light like a chip of glass, the chips moving in a broad, loose U. A separate skein of geese followed behind the U, and the lines undulated slowly, flickering on the clear sky. Migrating snow geese fly at an average speed of fifty miles an hour, usually at altitudes between 2,000 and 3,000 feet, though pilots have reported them at 11,000 and 12,000 feet and even once, over Louisiana, at 20,000 feet.

'Are those them?' Eleanor asked. 'Are those your geese?'

'Yeah, those are snow geese,' Matthew said. 'You see the way they're waving? That kind of gives it away.'

'Is that north?' I asked.

'That's due north,' Matthew said. 'That's Canada.'

They were on their way, heading for breeding grounds. It thrilled me, seeing geese. We turned our attention back to the house, but I kept looking up, hoping to see more geese. But the only birds visible were the vultures.

'This is it,' Matthew said. 'What do you think?'

'It looks like a house,' Eleanor said.

'Doesn't it?' Matthew enthused. 'I think it does. It's actually a house. I'll show you around.'

The driveway was covered with a mulch of shredded cedar bark. The air smelled resinous and biblical from the bark and crushed cedar needles. The mast gave off a low electric hum.

'The mast doesn't bother me,' Matthew said.

'But it sways,' said Eleanor, as though finishing his sentence.

'Yeah, it sways. It sways a lot in strong winds. It sounds like white water. It sounds like river rapids. In storms, lightning strikes it. I think of it as a lightning conductor. Lightning's never going to hit the house; it'll hit the mast.'

We stepped up on to the squared-off concrete foundation and entered the house through the gap of a side door. Trestles, door jambs, window casements and miscellaneous beams were strewn about the bare concrete. Matthew walked us round the site, describing the layout, furnishing and atmosphere of rooms. He saw the finished home in his mind's eye; he envisaged colours, materials, textures, designated zones, the workings of light, the logic of passage and enclosure; he understood how the house, his habitat, would accommodate his habits and inclinations. One bathroom would be left part open to the elements, with a free-standing clawfoot tub and a view of the mast's red lights, and stars beyond them.

'I'm the only one who knows what it's going to look like,' Matthew said. 'Right here I'm like the tiny God.'

'Have you been living in the tent?' asked Eleanor.

'Just for the last few days. It's been warm enough.'

He'd pitched a tent at the edge of the clearing: rugged, off-white canvas on a pine frame. The tent had the character of a tabernacle next to the cedars. Inside, on a wooden platform, was a single bed covered with a charcoal-grey blanket and a green sleeping bag unzipped to the waist, a mantle lamp hanging overhead – a glass chimney with voluptuous throat and swell and a sooty streak up one side where the flame had licked. One end of a table-top rested on a crossbeam of the tent frame, the other on a twisted cedar branch, making a desk surface for *Fine Homebuilding* and *Natural Home* magazines, half-drunk mugs of coffee, rolls of string and masking tape, and sheets of lined yellow paper filled with sketches, projections, floor plans and bullet-pointed lists. Two photographs were taped to the tent frame: a band of elephants strolling through lush Kenyan grass, and a young woman at the foot of a ski-slope, her teeth the same gleaming white as the snow, the photographer himself reflected in both lenses of her sunglasses.

A path led from the tent to the front door of the house, and to the left of the door stood a small statue, carved in pink stone: a bearded man wearing a robe cinched at the waist by a cord, holding out a bowl that the sculptor intended to be a bird bath. Matthew had bought the statue from a finca in Mexico.

'Who is it?' I asked.

'Saint Francis,' he answered.

Just then I felt Eleanor's hand on my left shoulder. I turned round; she handed me a feather.

'I want to give you this,' she said. 'I guess it's a hawk's.'

A large feather, chestnut brown, crossed near the tip by a single black bar: a flight feather from the wing or tail of a red-tailed hawk, much larger than the contour feathers that cover the body, and with a slight curve to its shaft, or rachis. Its vanes weren't symmetrical – typical of flight feathers, which have a narrow outer vane to cut the air. Each vane was made up of hundreds of fine barbs, branching out from the long central rachis, the shaft that thickened to a hollow quill, or calamus which would have anchored the feather in a follicle under the bird's skin. With my fingers I teased the barbs apart, opening up a section of the vane. Each barb itself resembled a feather, with rows of cilia called barbules branching from a hair-thin shaft, or ramus, and I knew that along each barbule, too small for the naked eye, there were tiny projecting hooklets called barbicels that hooked into the barbules and barbicels of adjacent barbs, meshing together, zipping one barb to the next, forming the flexible, continuous, almost-woven fabric of the vane. The feather's rust-brown colour was produced by melanin pigments concentrated in the barbs. Close to the calamus, the barbs became white and fluffy, like down feathers.

I held it in my right, writing hand, quill gripped between thumb and forefinger, resting on the groove of the middle finger, the rachis curving back over my wrist.

'Keep it,' Eleanor said.

I put the hawk's feather in my shirt pocket, quill first, and we stepped through the doorspace, across the concrete threshold, into the house. Matthew was holding a measuring tape against a piece of pine, marking off lengths with a pencil, sun glaring off the concrete planes.

We all heard the belchy bass chugging of a digger approaching through the cedars.

'Here's Mr Harper,' Matthew said. He thumbed a button – the metal tape rattled back into its palm-sized box – then dropped the tape and pencil into one of his belt pouches. Eleanor and I followed him out of the house as a yellow Caterpillar digger entered the clearing, its driver invisible behind tinted black windows. The driver lowered the loader and set to extending the driveway, pushing earth and rubble into ridges, dust rising around the machinery.

'This guy's the best dig-truck driver I've ever seen,' Matthew shouted. 'Those teeth at the front end? He uses them like fingers. Like his own fingers, it's so delicate.'

The two turkey vultures were still soaring overhead, hanging on updrafts created by wind deflecting off the hills, or on thermal columns rising off roads and clearings in the cedar forest. Their wings were held upwards in flat Vs, the slotted feathers at each wing-tip curving upwards. The vultures sometimes rolled from side to side, riding the gusts and currents, but I never saw them flap their wings. Their gliding was poised and effortless: weight, lift, drag and thrust – the forces essential for flight – were in perfect balance. Now and again the shadows of vultures slid across the concrete planes.

The yellow Cat chugged up a track into the cedars, then reappeared with a smooth-sided limestone slab in its raised, toothed trough. Matthew wanted his driveway to loop around a cedar, the only tree left standing on the building site. Mr Harper drove towards the cedar, stopped the Cat and lowered the trough, and the limestone slab tumbled over the teeth like an old tomb, coming to rest on the rubble and shredded bark, raising a cloud of dust.

Matthew wanted the stone to be in just the right place, and Eleanor and I stood a few steps back as he communicated with Mr Harper by means of an antic semaphore, a repertoire of pointing and waving gestures, and Mr Harper used the trough to nudge and coax the slab according to these instructions. Matthew devoted all his attention to the placing of the stone. He was building a house. Trees, windows, doorways, statues – these were to be his life's fixed marks, unchanging and dependable, his points of reference. He had to get it right. Slowly, in line with Matthew's signals, Mr Harper worked the stone over to its allotted place next to the cedar.

'That's it!' Matthew shouted. 'Right there!'

The Cat drew back from the stone, tracks clanking. stood with his hands on his hips, nodding with approval. He looked at the stone, then at the house, the canvas tent, the red and white mast, as if contemplating the distribution of landMatthew marks, registering just what was here and what wasn't. The vultures kept gliding round and round, birds circling on a child's mobile.

'OK?' Eleanor yelled at Matthew.

'Yeah,' he shouted. 'Looks good.'

We got back into the Mercedes. Eleanor drove across the clearing towards the opening in the cedars, and I looked back over my shoulder at the house, the mast towering above it, Matthew striding towards the Cat, towards Mr Harper. The cab's door opened as Matthew approached, but cedars blocked my view from the car the instant before a figure emerged.

*

AFTER SUPPER, which we ate in the kitchen, sitting on stools at the sideboard, Eleanor filled the larger of the two skillets with water, and set it to boil on the glowing electric hob. She was wearing a pink sweatshirt pinned with a brooch: a gold harp with four short strings. She transferred leftovers to simple china bowls, covered the bowls with tinfoil and put them in the fridge, which was already full of such bowls, the fridge-light shining off their foil skins. She went back to the cooker and stared down at the water, searching it for bubbles.

In the living-room she'd drawn the curtains across the sliding glass doors to the balcony. Brass wall fixtures held lights resembling candles: small, flame-shaped bulbs of opaque white glass. The deep, warm tones of the walnut were comforting. I sat on the mulberry sofa; Eleanor sat in her leather armchair. She rummaged in a cloth bag lying close to her feet and pulled out a piece of unfinished tapestry, wool-ends hanging loose off a square of white gauze. It was to be a cushion cover depicting two rabbits. She put on a pair of glasses with translucent, blue-tinged plastic frames, reached out for the Anglepoise lamp with her right hand, directed its light like a dentist, and set to work.

'So what I want to know is, where is this all going to end up?' she asked, not looking up from the tapestry.

'On Baffin Island,' I said. 'Right up in the Canadian Arctic.'

'Why Baffin Island?'

'I read that the largest concentrations of geese nest on Baffin Island,' I said. 'There's a train from Winnipeg to Churchill, on Hudson Bay. You can fly over Hudson Bay to Cape Dorset, at the south-west tip of Baffin. From Cape Dorset I'll try to get out into Foxe Land to see snow geese.'

Foxe Land was named for Captain Luke Foxe, who'd sailed from England in May 1631, hoping to discover the North-West Passage, a navigable sea route round the north coast of North America to Japan, China and India. His ship, the *Charles*, a pinnace of seventy or eighty tons burden, had a crew of twenty men, two boys and a dog. Foxe kept a journal of the voyage, later published as *The North-West Fox*, and I'd relished its descriptions of billows, races, overfalls and flood tides, and mild, calm days when pilot whales fluked and sounded just off the bow in a 'sea so smooth as if it had been made ready to bowl upon'. Approaching Hudson Strait, the pinnace encountered drifts of floe fragments and freshwater bergs calved off glaciers, and in Hudson Bay Foxe saw white beluga whales, polar bears swimming from floe piece to floe piece, dramatic auroral displays, and a sea unicorn or narwhal, 'his side dappled purely with white and black; his belly all milk-white; his shape, from his gills to his tail, fully like a mackerel; his head like to a lobster, whereout the forepart grew forth his twined horn, above six foot long, all black save the tip'. The *Charles* sailed round the south-west tip of Baffin before returning to England, and Foxe named Cape Dorset in honour of his sponsor, Edward Sackville, Earl of Dorset.

On 26 August 1631, he saw geese flying south over Hudson Bay: 'A N.N.W. wind,' he wrote, 'hath conveyed away abundance of wild geese by us; they breed here towards the N. in those wildernesses. There are infinite numbers, and, when their young be fledged, they fly southwards to winter in a warmer country.' Auroras, narwhals, wildernesses, infinite numbers: my restlessness, my appetite for snow geese, grew stronger line by line. In May 1929, almost 300 years after Foxe's voyage, the Canadian ornithologist John Dewey Soper set out from Cape Dorset to search for the breeding grounds of the snow goose. With two Inuit assistants, Kavivow and Ashoona, Soper established a camp just north of Bowman Bay, and during the first half of June he watched waves of geese pass overhead, obeying their 'furious northern urge'. On 26 June the three men found their first nests. 'The long quest,' Soper wrote, 'had ended.'

'In about three months' time, I hope I'll be in Foxe Land,' I told Eleanor, who had been switching her attention between tapestry and visitor with impressive, quiet acuity.

'Well, good luck to you,' she said. She smiled; her crow's feet deepened.

*

STREETLIGHT FILTERED through the blinds into the small wood-walled guest room, picking out the lines of the birdcage on the chest-of-drawers, the curving fretwork culminating at the zenith ring. I tended to wake up early. In the months I'd spent at home, waiting for my strength to return, I'd wake when it was still dark and lie in the small bed, across the sag in the unsprung horsehair mattress, interrogating the chain of events, imagining the life I was missing, fearing further setbacks, my mind grinding its teeth. That room's curtains had not changed since childhood: they were blue, with parallel bands of giraffes, lions, elephants and apes in browns and dark greys marching from left to right, as if to arks. When you drew the curtains back the animals huddled together in the enclaves of the pleats and folds, and the fact that they were there, and just as I remembered them to be – the fact that *now* was agreeing with *then* – was itself reassuring: a conduit to less equivocal days, a mark of steadiness in the chaos of illness and its treatments. Light gathered in the bands of animals and the intervening blocks of blue, and slowly the shape of Everest emerged, with the biplane hanging like a toy far below the level of the summit. The curtains, the picture, a simple chair with my father's jacket draped on its shoulders – these objects filled particular vacancies as if designed for them.

Impossible, on such mornings, to imagine that one day I would be in Austin, Texas, on my way towards the Canadian Arctic in the company of a bunch of birds. My early waking at that time may have been a sign of depression, but it persisted as a habit even after the crisis had passed and the strength of my anxieties had waned. So I woke up early in Eleanor's house, in the room that had been Matthew's room, and there it was, the birdcage standing on the chest-of-drawers, birdless, gilded even in the half-light, its lines concluding as if in a knot at the zenith of the dome, my glance alighting as no bird could on the dowel rod that threaded the cage, and then on the drawing of the cowboy pitched in the air, his hand on the reins and his feet in the stirrups his only points of connection to the bucking horse.

I hadn't intended to stay more than a night. But Eleanor encouraged me to settle in, to make myself at home. She taught me some of the house's quirks – the special expertise required for shutting off a tap; the chest where blankets were kept, pulled two inches out from the wall so the lid's back edge didn't scrape on the panelling when you opened it – and I learned others for myself, like the way the puck spun off the wind-chimes, the noise the slatted doors made as they swung to a shut in gradually shorter arcs behind you. I sat at the glass-topped table on the balcony, reading and writing, learning about birds, at ease, finding my feet in America. Several days passed before descriptions of Zugunruhe reminded me of my purpose, and restlessness took hold again. I hadn't come to settle in. Spring was under way, we were well into March, snow geese were pressing towards Winnipeg. I had to get going.

Eleanor slid back the glass doors and stepped out on to the balcony.

'What do you think about going to see the bats?' she asked.

She said that bats arrived from Mexico each spring to roost under a bridge downtown. You could see them at sunset, when they came out to forage for insects.

I told her I loved the idea, but I'd been thinking about moving on.

'You're worrying about those birds. They're leaving you behind, right?'

'I've got some catching up to do.'

'I guess those geese could be in North Dakota. I guess we better get you on the Greyhound. We better rush you up to Fargo!'

*

THAT EVENING, we drove to see bats. Eleanor parked the caramel Mercedes close to the river. She was wearing a navy blue anorak with a chunky white zip up the front, and her white hair seemed oddly contingent on the zip, like a bloom at the tip of a stalk. We walked out across the Colorado River on Congress Avenue Bridge, stopping at the midpoint of the span, looking north up the avenue to the domed capitol, and west downstream towards another bridge and the cedar-covered hills far beyond it. A small plane flew across the hills, trailing a banner whose commercial message or surprising essential truth wasn't quite legible in this fading light.

'This is a good spot, right here,' Eleanor said. We stood at the balustrade, looking east over Town Lake. 'We've got bats right under our feet.'

In 1980, reconstruction work created expansion joints in the deck of the Congress Avenue Bridge – parallel inch-wide grooves, each more than a foot deep, which Mexican free-tailed bats soon took to roosting in, the temperature and humidity just right for raising pups. Four or five inches long, dark grey-brown, with big forward-pointing ears and wrinkles on their lips along the muzzle, the freetails winter in Mexican caves and return to Austin each spring. Like birds, bats have internal circadian and circannual clocks, entrained to the natural year by Zeitgebers, triggering migratory behaviour at appropriate times.

'I haven't come down here to watch bats for I don't know how long,' Eleanor said.

People were gathering on the bridge and in the small public park on the south side of the river, sitting on rugs on the grass bank, taking up positions either side of us at the balustrade. A boy wearing a plastic black Batman cape argued with a girl in a bat's ear hairband. Two women led Great Danes along the riverside path, the dogs attired like conscientious bicyclists in fluorescent yellow collars and sashes. An oarsman moved silently beneath us, sculling upstream, feathering the blades as if they were his own palms held above the water, his seat rolling back and forth on greased coasters and rails, the dips of the oars leaving a series of paired circles like a shoe's eyelets in his wake. We felt the shudder of the asphalt panels as traffic crossed the river on Congress Avenue.

'Those bats'll be coming out any minute,' Eleanor said.

Spars of red and white shimmer lay across the water, thrown off the city lights. Everyone was waiting. A dark blue sedan pulled up behind us, pausing on the bridge just long enough for an old man to step gingerly on to the kerb, assisted by a male nurse dressed in a white, dog-collared hospital tunic. The old man wore a dressing-gown over a green hospital smock, and the thin shins visible below the hems of these robes were sleeved in the white compression stockings that prevent deep vein thrombosis in the bed-bound. He wore bright purple slippers with pineapples surprisingly embroidered on their topsides, and his face was gaunt, pared of all substance, with cheekbones showing like stanchions under pink, brittle-looking skin, and a tuft of white hair like a wisp of smoke off his scalp. He moved shakily with anxious inch-long steps to the edge of the bridge and took his place at the balustrade to my left.

'How are you holding up, Mr Mitchell?' asked the nurse, who had short black hair.

'Pretty good, I guess,' said the old man in a weak, high, trembling voice.

He was just in time. Without warning, bats began dropping from the grooves under our feet, streaming past the live oaks and cypresses of the southern shore. People said 'Oh!' and 'Ah!' as if at fireworks as the freetails accelerated away from the bridge, a tube of shadow sloping upwards into blue-grey light, vibrant with points of agitated air – as if the bridge were sighing them out, a gasp of breath in which each atom was figured by a bat. Their wings made a papery flutter, the rapid soft flutter of banknotes hurrying through a counting machine, twenty-five notes per second, and I tried to imagine beneath the flutter the click-din of echolocation, the drumfire of ultrasound pulses that allowed bats hurtling at a cypress tree to hear the fact of it and bear away, upriver.

During the Second World War, the US Armed Forces developed a scheme known as Project X-Ray, in which large numbers of Mexican free-tailed bats were fitted with small incendiary bombs, attached to their bellies by a short string and a surgical clip. The idea was that cages of bats would be parachuted over enemy territory and open at a particular altitude, releasing teams of explosive bats that would quickly disperse to buildings in the immediate vicinity. Once in their roosts, the bats would chew through the strings and release the bombs. But the scheme was dropped because the bats never dispersed; they stuck together, gathering at one or two roost sites. On one occasion hundreds of armed Mexican free-tailed bats escaped their test range in the south-western desert and blew up several military buildings and an elevated gas tank in a nearby town.

The last bats dropped from the grooves beneath us, unencumbered.

'There they go,' said Mr Mitchell.

'We didn't get here a moment too soon,' said the nurse.

'I told you, didn't I? Didn't I tell you?' His bony pink hands gripped the balustrade; his whole body was shaking.

'Yes, you did.'

Eleanor didn't take her eyes off the bats. The stream disappeared into dim light, a thick rope pulled by a stevedore. Mr Mitchell, truant from his ward, stood trembling at the balustrade. Eleanor's white hair was faintly luminous. Red lights glimmered on radio masts in the hills. The statue of Saint Francis was in place beside the door, the limestone slab beside the cedar tree. Birds were flying north according to inherited programmes. Cars passed back and forth on Congress Avenue. The crowd began to disperse, heading home. The dark blue sedan appeared on the bridge for the second time, and the nurse placed his hand gently on Mr Mitchell's head as the old man stooped into the open door.

3 : GREYHOUND

THE PROSPECT OF moving north across America as spring itself was moving north and millions of migrant birds were moving north with the warmth to their breeding grounds as the North Pole tilted gradually closer to the sun – this prospect was so exciting that when Eleanor knocked on the door before dawn, with the lines of the birdcage picked out by the streetlight, I almost jumped out of bed. Electric light jarred off the white-tiled kitchen floor and glinted off the piano magnets and off the corners and hinges of the fridge, and when Eleanor opened the fridge to get some milk I saw all the bowls with their tight foil skins like a range of drums – a set of small, tuned timpani on the white racks. We made tea, jigging the strings, and then I fetched my bag from the wood-dark bedroom, breezing through the slatted swinging doors, hearing them *thwup thwup thwup* to a halt behind me. Eleanor was waiting in the living-room, one hand on the table of tortoises, the other sprucing up her downy white hair. We drove through Austin to the Greyhound terminal, yawning in canon, and Eleanor parked the Mercedes at the entrance to the terminal building.

'Have a good time,' she said.

'Thanks for everything.'

'Don't mention it. Say hello to the geese for me.'

I watched the old caramel Mercedes leave the parking lot, *There's No Place Like Narnia* disappearing in the traffic stream.

Buses were berthed on the far side of the terminal: a fleet of silver-styled Americruisers basking under arc lights, decked out in sprinting blue greyhounds with thin, tapering snouts you could clasp in your hand like ice-cream cones. Automatic doors opened on a waiting area with bare beige floors and the featureless walls of transit zones, and luggage heaps, sleeping figures, illuminated vending machines, and ranks of screwed-down seating units, some of them fitted with small, coin-fired televisions in moulded black plastic casings. Passengers were standing around, waiting for gates to be called, checking their watches, wandering from one spot to another, carrying tubes of Pringles potato snacks, portable stereos, transparent Ziploc bags of cookies or muffins, black refuse sacks bulging with laundry, rolled-up sleeping-bags, green Army Surplus kitbags, suit bags, duffel bags, rucksacks, pillows, cooler boxes, comfort blankets and swaddled, sleeping babies.

The Greyhound left Austin at seven o'clock in the morning. The schedule that came with my ticket told me I would arrive in Fargo, North Dakota, at twenty to six the following afternoon, having changed coaches in Dallas, Oklahoma City, Kansas City, and Minneapolis. Fargo was about 1,000 miles due north of Austin; snow geese would be flying due north from Texas to North Dakota, as if along the lines of longitude. *North* – the word had amplitude now, as if all possible destinations and endings were gathered up in it. North trumped all other places. I wanted the front seat in the Greyhound for its widescreen view of the north, but it was taken, so I sat two rows back on the right-hand side, across the aisle from a young woman and her son, a boy of five or six with brown hair cut in a severe fringe, like a monk's tonsure, and a book of join-the-dots pictures to which he applied himself with a monk's diligence, working with felt-tip pens in the dim light, conjuring motorcycles, tennis racquets, kitchen blenders and giraffes from strews of black points while his mother, who had spiky blond hair and a thin, bony face, laid her head against the window and slept despite its shuddering.

A storm broke as we left Austin. Thunder rolled; lightning lit the rooftops in quick, jagged brightnesses; water sluiced across the Greyhound's windscreen, batted left and right by the long, jointed wipers. I thought about Matthew in his tent at the edge of the cedars, hard rain thrumming on the off-white canvas as bolts struck the tall mast. Then the storm passed, the sun rose somewhere beyond Baton Rouge, and the Greyhound surged towards Dallas on Interstate 35. I gazed blankly through the tinted window, lulled by the hum of wheels on even asphalt panels, with flat country skirting past on the far edge of my attention – a Texas bric-a-brac of motels, outlet malls, dancehalls, subcourthouses, pet-grooming salons, ministorages, swags of electric and telephone cables, a non-stop barrage of exclamatory hoardings and signs like heraldic shields raised high on steel masts, flashing the names of gas stations and franchise restaurants as the coach coasted past Georgetown, Temple, Waco and Italy, 200 miles north to Dallas.

I boarded a new bus in Dallas, its driver a tall, lean, narrow man, like a cigarette dressed in the grey Greyhound uniform, with sleeves rolled neatly to the elbows and silver hair cut short at the back and sides, swept back on top and glossed with brilliantine. He wore a brown leather belt embossed with an eagle, laterally extended, and a dated-looking digital watch with a calculator keypad underneath its scratched display. He sucked on a toothpick, smoothed his hair back with both palms simultaneously, and addressed his passengers as 'folks'.

'Now please remember, folks,' he said into his microphone, lips brushing the metal mesh as we proceeded through the Dallas suburbs, 'that we do have ladies and children on board. So let me say, folks, that we do not want to say or do anything that would embarrass those good folks. No bad language. No lewdness of any kind. Now this may not pertain to you, but I'm saying it all the same. I'm deadly graveyard serious on this matter, folks.'

The folks in my vicinity included, in the front seat, a frail, white-haired lady wearing a denim shirt adorned with homemade four-pointed appliqué stars and a smiling brown crescent moon. She tried repeatedly to engage the driver in conversation, but he didn't respond; he was deadly graveyard serious on all matters of road safety. Behind me, a younger, dark-featured woman, dressed in an orange tracksuit, had immersed herself in a paperback entitled *Blues for Silk Garcia*, and across the aisle sat a burly man with a ponytail falling across his chest, black hair fanning out on a T-shirt that read, *Since I Gave Up Hope, I Got Much Better*. Behind him, through a gap between headrests, I glimpsed a woman in small, round, wire-framed spectacles, stroking the head of a sleeping baby.

The toothpick clamped in the driver's teeth pointed north up Interstate 35 like a compass needle. The coach hardly wavered from its cruising speed. Freightliner, Eagle and Kenworth rigs with sleeping cabins and gleaming silver chimneys drew level with the Greyhound, then accelerated past, hauling Utility, Stoughton and Great Dane freight containers. There were other Greyhound and Jefferson Lines coaches in the current of the highway; recreational vehicles with the names Jamboree, Chieftain, Prowler and Nomad splashed on their creamy white foreheads and mountain bikes lashed to their backs; state trooper cars with four trunk aerials bending backwards like grasses in the apparent wind; entire prefabricated houses proceeding with due caution along the inside lane; and station wagons, trailers, vans, jeeps, pickups, hatchbacks, sedans – the hard c sounds of American traffic: Mack, Cadillac, Pontiac, Camry, Buick – with mottoes (*Grace Happens!*) on rear fenders, and dogs leaning from open windows, nosing the windspeed. Stars and Stripes of immaculate parachute silk rippled at the gates of lots and salerooms. A great blue heron lifted from a marsh. A flock of ten or twelve ducks flew alongside us in compact formation, a tiny clutch of the millions of birds that were moving towards Canada with the spring, subject to circannual rhythms and Zugunruhe, far outnumbering the people in vehicles passing Denton, Gainesville, Marietta and Ardmore on their way to Oklahoma City and all points north.

These birds possessed compasses as well as clocks. In 1949 the German ornithologist Gustav Kramer had observed young migrant starlings in an outdoor aviary. Kramer was interested in their ability to navigate. 'Such a conspicuous phenomenon as the long-distance flights of birds,' he wrote, 'has profoundly penetrated into man's consciousness, and it is a very simple further step to ask how they find their way.' At the end of the summer, Kramer's starlings, which came from the Baltic region, exhibited 'a distinct tendency to migrate south-west'.

The following year Kramer transferred these birds to circular pavilions in which vision was limited to six windows, distributed symmetrically round the compass, with landmarks carefully excluded from view. Mirrors were mounted at each of the windows, reflecting sunlight into the cages at ninety-degree angles. The drum-shaped pavilions rested on transparent Plexiglass bases: observers lay underneath, looking up at the birds, recording their behaviour.

The starlings displayed Zugunruhe at the appropriate time, with a tendency to hop towards the north-east, the appropriate direction for spring migration. Then, by manipulating the mirrors, Kramer changed the apparent direction of the sunlight. The starlings changed direction accordingly: the birds were using a sun compass. Such a mechanism, Kramer noted, could not be effective without an internal clock. The sun's position relative to a point on the Earth changes by 15 degrees every hour: the starlings must have some way of compensating for this apparent movement. 'The migratory activity on some days lasted for six hours,' Kramer wrote, 'from the early morning until noon, which corresponds to a movement of the sun through about 90 degrees; yet the bird's direction remained unaltered.' He christened one of the starlings Heliotrope, like the flower, from the Greek for 'tending towards the sun'.

The discovery of the sun compass was a first step towards answering Kramer's question: how do birds find their way? But many birds, including starlings, are able to migrate on cloudy days when the sun is hidden, and many birds migrate at night. The sun on its own was not enough. Birds must possess some other means of orientation.

In the terminal at Oklahoma City the Greyhound slogan – *Where Can We Take You?* – was printed on banners hanging above the screwed-down seats, the sprinting greyhound trade-mark a cartoon of speed, efficiency and kinetic grace. Coaches pulled up outside, their front-ends sinking on hydraulic mecha-nisms to the kerb, like camels kneeling. It was cold now: in just a few hours we had outstripped the spring. The terminal was another limbo, an in-between place, a corral for itinerants, with nothing to mark it out as here, not there. Travellers attended luggage heaps or loitered by a snack bar where helical ribbons of yellow Victor flycatcher hung from the ceiling and an old fan on a white stand turned from side to side as if watching a very slow game of tennis, or they leaned over the lights and bubbling electronic music of Addams Family pinball, or gazed into video games, piloting Spitfire, Zero and Shinden fighter planes through the puffs of digital flak and pixellated gunfire streams of *Strikers 1945*, and working the wheel and pedals of *Cruisin' USA*, accel-erating through redwood forests, across piñon-dotted Nevada deserts, over the Golden Gate Bridge, along the Florida Keys, up scenic Rocky Mountain passes and down broad Manhat-tan avenues, skipping from state to state, a primer of America flashing in the corner by the vending machines.

People milled about, waiting to board buses or greet other

people. Two women in their sixties, sisters, wearing long pleated skirts and hand-knitted cardigans, with salt-and-pepper hair and spectacles hanging from their necks on colourful braided strings, celebrated their reunion in the temperature-controlled terminal by placing their hands on each other's shoulders and delivering prim, delicate kisses to both cheeks like champagne glasses clinked together in a toast. A boy gazed up at a wall of gunmetal luggage lockers in three sizes, corresponding to handbags, over-night bags and suitcases, and then went right along the bottom row, trying the doors of the largest lockers. At last, one opened: a red-tagged key was still inserted above the coin slot. The boy looked around. He was planning something and didn't want to be observed. He climbed into the empty locker and pulled the door shut behind him. A minute or two later, a man approached the wall of lockers. He was in his late thirties, with a pale face and black hair so neatly parted the division resembled a chalked line. He wore a blue suit, a red-striped shirt open at the neck, a white T-shirt underneath; he carried a brown leather suitcase. He put the case down and scanned the rows of lockers. He noticed the key with the red tag. He opened the locker. The boy was waiting inside, on all fours, and as soon as the door opened he stretched his head out, a creature emerging from its den, beaming up at the man in the blue suit, who took a step back, bewildered but not alarmed, as though this apparition were merely a trick of his fatigue, the kind of thing you should expect in the fooling of these distances.

The bus for Kansas City was announced. The white-haired lady in the denim shirt decorated with the moon bagged the front seat. I sat behind her. The coach filled up: a man in torn jeans carrying a guitar, with the sections of a fishing-rod taped to its neck, the reel resting on the strings above the sound-hole; a girl hugging a lever-arch college file against her chest; two Amish elders dressed in black and white, with long grey beards and the

stern countenances of patriarchs; a man in a dark grey overcoat
holding a new cardboard box marked 'Stetson' on all sides: a
promise of pure hat. Just before we were due to depart, a woman
climbed the stairs and stood in the aisle, scanning the seats for
vacancies. She was Eleanor's age, with a thick, upswept crown
of silver-grey hair, a black canvas tote bag hanging from one
shoulder. She was ablaze with primary colours, dressed in a
yellow polo-neck, faded blue jeans, white socks in blue leather
sandals, large plastic-rimmed spectacles, and a bright red sleeveless fleece vest pinned with a badge that said, *Women Are Not
Born Republican, Democrat or Yesterday.* She took two steps
forwards, paused, looked the coach up and down, then plumped
for the seat next to mine, setting the tote bag on the floor
between her blue sandals.

'My name is Jean,' she said, holding out her hand. 'Pleased
to meet you.'

Our driver was short, with thin, mousy hair, and heavily
built, his grey shirt bulging like a laundry bag. He spoke into a
microphone as the coach drew out of Oklahoma City, warning
all passengers that the consumption of tobacco, alcohol and other
intoxicants would not be tolerated.

'Anyone who wishes to make use of personal stereo machines,'
he continued, 'I'd ask you to please first hold the headphones
out at arm's length, or as high as you can above your head, and
if you can hear a noise then spare a thought that your neighbour
will hear it also. And for anyone taking care of children on
board, keep them in their seats now, because if I have to swerve
or stop all of a sudden that little boy or girl is sure going to fly.'

Late afternoon. The Greyhound continued on Interstate 35,
crossing from Oklahoma into Kansas. Trees, signs and telegraph
poles spun past; we pushed north on the humdrum basis of
vibration and engine hum. I kept looking out for snow geese.
I kept thinking of the birds from Eagle Lake, imagining that

they were flying overhead, thinking that if only I could lean from the window and get a clear view upwards I would see them. Jean's hands were resting on her thighs. She was tanned; the pink frames of her glasses had miniature rococo scrolls at their corners; now and again she looked anxiously behind her, down the length of the bus, as if expecting to see somebody she knew. She had a brightly-painted watch with Adam and Eve represented on the strap's two sections, a large red love heart behind the hands, and smaller hearts at three, six, nine and twelve o'clock.

'Have you come a long way?' she asked.

'From Austin.'

The seats weren't spacious. Strangers talking to each other for the first time do not normally hold their heads so close together, and I wasn't sure if I should turn to my left in order to look at Jean, or look straight ahead, or continue gazing from the window as our conversation proceeded. We both spoke quietly, almost in whispers, as if to keep what we said confined to the immediate, enclosed space of the double seat. Although there was no screen between us, and no penance to be done when business was concluded, our conversation bore a trace of the hushed, boxed-in disclosures of confessionals.

'How far are you going?' I asked.

'Minneapolis. You?'

'Minneapolis, too. Then Fargo.'

'Are you visiting relatives? My sister's in the hospital in Minneapolis.'

'I'm going to look for birds.'

'Which birds?'

'Snow geese.'

'That's interesting. I don't know much about those. We live in the city.'

She lived in Oklahoma City but had grown up in New Orleans. The sky was darkening, restoring powers of reflection to the window glass, the sun low on the far side of the Great Plains. The Greyhound was still heading due north, and sometimes, while Jean and I were talking, I savoured the clarity of this direction, as if Interstate 35 were a line to which migrating birds could cleave as they travelled from wintering to breeding grounds.

A sun compass on its own was not enough: birds had to possess some other means of finding their way. The idea that organisms might use the Earth's magnetic field for orientation was first proposed in the nineteenth century but not taken seriously by ornithologists until the 1960s. Magnetic field lines leave the Earth at the south magnetic pole and enter it again at the north magnetic pole. In between, these lines form varying angles of inclination with the horizontal: ninety degrees at the poles, zero degrees at the equator, changing systematically as they span the globe. The magnetic field provides a gradient map which could, in theory, be a source of reference for migratory birds.

Wolfgang and Roswitha Wiltschko kept more than 200 European robins in octagonal wood and plastic test cages from which all visual clues were excluded. Each cage contained eight perches, one for each side of the cage, and each perch was connected to a microswitch that produced a signal when the bird hopped on it. Robins are partial migrants: some migrate to Mediterranean and North African winter grounds, while others are European residents all year long.

The Wiltschkos screened off the Earth's magnetic field with a steel vault and recreated it artificially by means of wire coils carrying electric currents, known as Helmholtz coils. In the spring, their robins began to display Zugunruhe: whirring their wings, hopping, flitting from floor to perch. The microswitches recorded the birds' directional tendencies: the robins were trying to fly north. When the direction of the experimental magnetic field was shifted, the robins changed direction accordingly. Even when magnetic north was shifted to geographic south, the robins followed suit, flying directly away from their appropriate destinations.

'The direction the birds take for "north",' the Wiltschkos concluded, 'does not depend on the polarity of the magnetic field.' Their robins seemed to be referring instead to the inclination angles of the field lines. In the spring, robins flew in whichever direction the inclinations became steeper, because this meant they were flying towards the pole. In the autumn, they flew in whichever direction the inclinations flattened out, because this meant they were flying towards the equator. The polarity of the Earth's magnetic field has reversed thirty times over the past 5 million years: the Wiltschkos noted that a compass which depended on inclination angles rather than polarity would not be affected by such switches.

'We lived on Music Street,' Jean confided, her voice languid with southern twang, 'in Gentilly, in New Orleans, me and my brother and sister, in a cramped little house with a backyard right up next to a railroad. The boxcars went clang-clang, and we rushed to the wall, and conductors smoking cigarettes waved at us from their cabooses. There was a washing-line in the yard and my mother hung everything on it – whites, delicates, you name it – and everything got covered in soot from the trains, all grimed up with coaldirt. My father left the house at two or three every morning to deliver milk for Mueller's Dairy in Elysian Fields, up and down Franklin Avenue, and from one o'clock to nine o'clock at night he drove a public bus. The milk truck had a freezer box on the back, and once, when we played hide and seek, I locked myself in it.'

'Did anyone find you?'

'No. When I felt the engine starting I banged on that door for all I was worth. I nearly got delivered to Elysian Fields! We played hide and seek a lot, and there was a craze for hula hoop. I loved hula hoop. Every kid in the neighbourhood loved hula hoop. We had our own hula hoop club. Hoops were fifty cents each at McCoy's dime store. We were quite boastful when we got a different colour – yellow, green, blue, you name it. Our dream was to acquire a hoop for every colour of the rainbow. We had small hoops to swirl on our arms at the same time as the big hoops. With a hula hoop, you don't circle the hips. It's a forward and back movement; it's getting a rhythm, and once you've got a rhythm, you hit it forward and back.' Jean was moving in her seat beside me, raising her arms, turning her shoulders from one side to the other, gyrating and shuffling, her own younger self roused inside her.

'Not bad!' I said.

'Oh, I was good!' Jean said, laughing. 'And you know what I was good at too?'

'What?'

'Tennis.'

'Really?'

'Oh, yes. My passion was tennis. We had a public tennis court on St Roch Avenue, in St James's Park. There was a baseball field, six swings, a slide and a public tennis court. The New Orleans Recreation Department sent a tennis pro around the neighbourhoods. They held raffles, tennis raffles. Tickets cost a quarter. The prize was you could win a racquet. I pleaded with my mother in the kitchen: "Can I have a quarter for the tennis raffle? Can I have a quarter for the tennis raffle?" I went on and on and in the end she said, "All right. I'll give you twenty-five cents. I'll give you a quarter for the tennis raffle." But there was a condition attached. Tickets were a letter of the alphabet and then a number from one to ten, and my mother said she'd only give me the quarter if I bought ticket J1, because that was my initial, J for Jean, and she thought that was going to be the lucky ticket. So I got it, I got J1, and they pulled J1 out of the hat, which is how I got my first ever tennis racquet.'

'You were destined to play tennis.'

'Exactly. I played tennis whenever I could. I loved it. But we were having a little trouble at home. After school my father locked me in my room. I heard him yelling at my mother and then he started to hit my mother and yell at her more and more. So I went to live with my sister in a boarding house. We had this funny old landlady with I don't know how many cats. I lived there until I finished high school. We had very little money. My sister washed hair to keep our heads above water. At the grocery store cans of vegetables were seven for a dollar, so each week we saved a dollar and ate one can of vegetables every day of the week.'

'What happened when you finished high school?'

'I'm going to tell you. I went to the convent. Not in New Orleans, but in St Louis. I became a sister. Something in me was saying, "The thing for you to do is be a sister." Recently, a woman who was a baton-twirler at the University of Texas asked me, "Aren't you embarrassed that you were a nun?" and I said, "Aren't you embarrassed that you spent all that time throwing a stick in the air?" I wasn't embarrassed. I knew it was something I had to do. I was clothed, fed and educated; I had friends; I got to help people. We spent a significant part of each day in silence and I believe that silence healed me. I learned discipline. We were taught to feel ourselves unworthy. We prostrated ourselves on the ground. If you broke something, there was a punishment. We had to lie flat on the ground with our arms extended like this and say Hail Marys for close to ever. We had to sleep with whatever we broke, just so we never forgot it. People were sleeping with fans, bowls, pots, plates, cups, you name it. Once I broke a statue of St Joseph. This statue was Mother Superior's favourite. It came from France. I had to sleep with the statue of

St Joseph and a lot of people said I was lucky just to have a man in my bed.'

We heard the indicator clicking; the driver manhandled the wide steering-wheel; the Greyhound turned off the Interstate. I'd got used to the bus making these stops, pausing at gas courts for short breaks between terminals. Passengers disembarked, lit cigarettes, sought restrooms, performed simple stretches, and made calls in hooded telephone booths, saying, 'Did you feed the fish?' or, 'See you tomorrow, Mushroom,' or, 'Don't ask me what I was doing, Marla! How should I know what I was doing?' Exhausted, far from home, we roamed the store aisles, fluorescent lights glaring on washed white-tile floors, the outdoor smells of exhaust fumes and gasoline mingling with the aromas of stale coffee, hot dogs, microwave burritos and drooling yellow cheese sauce squelched on to card boats of Gehl's tortilla chips. On the Greyhound you had only to gaze from the window and day-dream, but now you had to reckon with the siren-song of brand names in brimming racks – Chex, Dots, Runts, Twizzlers, Munchos, Rain-Blos, Lorna Doones – and weigh the merits of Dakota Kid sunflower seeds, Chupa Chups lollipops, Jack Link's Kippered Beefsteak Jerky, forty-four-ounce pails of Barq's 'Since 1898' Root Beer, and powdered, chocolate-dipped and honey-dipped donuts, French Twirl Donuts, Mickey Egg Fluff Donuts, Mrs Freshly's Creme Filled Gold Fingers, and Flaas Raspberry Bismarks. There were refrigerated cabinets stocked with sodas, iced teas, spring waters, juices and flavoured milks, and some-times there were elaborate pipe displays featuring Irvin S. Cobb's Corn Cobb pipes and 'pre-smoked' Dr Grabow filter pipes – the Riviera, Duke, Royal Duke, Omega and Savoy – made from imported briar, with standard bits and military bits; and Bryn Mawr Ream-N-Klean bristle pipe cleaners; and pouches of Black Cavendish, Gold Burley and Borkum Riff 'without a bite' tobaccos. We moved like sleepwalkers through this trove of nouns, drifting one by one back to the Americruiser. The door sighed shut when a complement was counted.

Jean and I settled down again in our seats towards the front of the coach on the right-hand side, the hands on her watch sweeping through the hearts. The Greyhound returned to the interstate, continuing north and north-east towards Kansas City. The driver spoke into his microphone.

'Your attention please,' he said. 'Now, I know you're going to think I'm going bald, but I found a hairpiece that belongs to one of you. I understand that you may be embarrassed to come up here and claim it, so I'm going to leave it at the front here and you can just wander up and claim it when we make the next stop and you get off the bus to smoke a cigarette or what have you. That's the end of my announcement. Thank you for your attention.'

We watched him lob the hairpiece at the windscreen. It slid down the glass and came to rest on the dashboard ledge, a curling, glamorous, brunette wig, tight-fitting, like a swimming-cap. It was dark now.

'Did you wear a habit?' I asked Jean.

'Oh yes. I wore a full white habit, starched and crisp. White because I was a novice. Black shoes, stockings, garter belt, gabardine petticoat, a white linen wimple covering my hair and both sides of my face, and a loose white scapular on my shoulders, hanging right down to my black shoes. What a business that was!'

'What about your tennis?'

'Oh, we had a court. There was a convent court. I played all the time. The tennis court was about the only place I felt at home. I remember one Sunday afternoon four tennis pros came to play a doubles game on our court. Every Sunday afternoon we'd have some kind of entertainment – someone would visit the convent to give a talk, or there'd be an activity of some sort or another. When I heard that these four tennis pros were going to visit I asked Mother Superior if there was any way I could get to hit some balls with them. That was so exciting to me, just the idea of playing with actual professional tennis players – you didn't get a chance like that every day of the week, and all I wanted to do was knock up for a few minutes, maybe have a game or two just for the experience of it. Anyway, Mother Superior looked at me for a while. She didn't say anything. I could see the wrath of God in her face and I thought that she was about to explode. She didn't say a word, she just stared at me, and I knew she was doing it to make me feel small. I started to apologize, but still she didn't say anything. I didn't know what to do. I looked around the room. I looked at my shoes. And then here it comes, oh it was the wrath of God, this terrible long lecture on my lack of humility and modesty and my arrogance. I was mortified. I felt truly shamed.'

'What happened?'

'Well, let me tell you. Sunday came around and the four professionals showed up, four lady tennis players in little white skirts and brand-new sneakers. All the sisters were sitting in the bleachers – there were three or four tiers next to the tennis court – all trussed up in stockings, garter belts, gabardine petticoats, wimples, scapulars, you name it. The pros started playing. They played a couple of sets. Mother Superior stood up by the net and made a speech, thanking the ladies for their exhibition. Then, I couldn't believe it, she said that Sister Jean-Marie was especially appreciative. Especially appreciative! One of the tennis pros said, 'Why doesn't she come and hit some tennis balls?' and another pro said, 'Sure, let her come and hit some tennis balls,' and of course this is in front of everybody, so Mother Superior doesn't really have a choice, she has to let me go and play with the professionals.

'I climb down off the bleachers and walk out on to the court, and remember I'm dressed in a habit and the pros are in these precious little tennis dresses, right up their butts, excuse me. One of them hands me her racquet and says she'll sit out so I can play. But I'm shaking. I'm on the forehand side. I'm receiving serve. I haven't even had a warm-up. I'm trying to stay low and concentrate on the server, who's bouncing the ball, preparing to serve. I could tell she wasn't going to go soft on me just because I was a nun, and sure enough the next thing I knew she sent down a firecracker, and I was so pent-up because I'd been so shamed that I hit it back with fury in it, I returned her serve as hard as I could, and the ball whizzed past the net person, straight down the line. It was a clear winner, no question. I'd watched the toss and seen just where she threw it. It came perfectly to my forehand and I made perfect contact, just as though the racquet were an extension of my body, and all the sisters in the bleachers jumped up and down and cheered. They whooped and hollered, and grabbed their scapulars, and shook them, and waved them in the air, their holy scapulars!'

It was getting late; people were sleeping. Reflections of red and yellow lights were sliding across the glass to my left, beyond Jean, then appearing unmediated on the near side, and other reflections were sliding across the glass between me and the lights, as if the Greyhound were revolving, or moving at the centre of revolving carousels of lights: streetlights and headlights, the red lights on rear fenders and radio masts, the winking red winglights of planes, hazes rising off the thick-sown lights of conurbations, the brightness of car lots (buffed hoods gleaming under Klieg lights), the neon fantasias of funfairs and casinos. The driver switched off the lights inside the bus, and stars were suddenly visible, constellations in the east: Hercules, Boötes, Virgo.

In the 1950s, the German ornithologist Franz Sauer suggested that birds might refer to the stars in order to determine their migratory direction. In the late 1960s, Stephen Emlen studied indigo buntings, a species that breeds throughout the eastern half of the United States and winters in the Bahamas, southern Mexico, and Central America south to Panama. Caged indigo buntings display intense nocturnal Zugunruhe in April and May, and again in September and October, the two periods during which their counterparts are migrating in the wild. When this restlessness began, Emlen placed his buntings in special circular cages: funnels of blotting paper mounted on ink pads and covered with clear plastic sheets. Birds in these cages could only see the sky overhead; all ground objects were blocked from view.

'A bunting in migratory condition,' Emlen wrote, 'stands in one place or turns slowly in a circle, its bill tilted upward and its wings partly spread and quivering rapidly. At frequent intervals the bird hops on to the sloping paper funnel, only to slide back and continue its pointing and quivering. Each hop from the ink pad leaves a black print on the paper. The accumulation of inked footprints provides a simple record of the bird's activity: they can later be counted and analyzed statistically.'

The buntings kept diaries: footprints lettered their seasonal restlessness.

Emlen put the cages outside on clear, moonless nights. In September and October, the buntings tended to hop south. In April and May, they tended to hop north. The cage walls screened the horizon from view: the birds could only see the sky. On cloudy, overcast nights, their orientation deteriorated significantly. Emlen hypothesized that buntings were able to determine their migratory directions from visual cues in the night sky.

Emlen then took his buntings into a planetarium. In September and October, using a Spitz Model B projector, he shone the normal autumn stars onto the dome. The buntings, appropriately, left footprints in the southern sectors of their cages. In April and May, Emlen projected the stars of a normal spring sky.

The buntings hopped north and north-east. But when Emlen switched off the projector and filled the dome with diffuse light, the buntings behaved just as they had done on cloudy nights outside: they were unable to determine their migratory direction. And when Emlen shifted Polaris to the east or west, the buntings changed their orientation to match the new 'north' or 'south', depending on the season.

To understand the significance of Polaris, the North Star, you first have to imagine that all the stars are fixed to a celestial sphere centred on the Earth. You have to imagine the axis on which the Earth is spinning. And then you have to follow the line of this axis from the North Pole up to the celestial sphere. The line intersects with the sphere at the north celestial pole, which happens to be very close to Polaris, a bright star located just off the tip of Ursa Minor, the Little Bear. Due to the rotation of the Earth, the celestial sphere appears to rotate clockwise around Polaris.

The axis of celestial rotation is always aligned with geographical north. Buntings, Emlen found, were determining direction by reference to the rotation of star patterns. The constellations move across the sky with an angular velocity of fifteen degrees an hour, but their shape remains constant, and each maintains a distinct relationship to the North Star. When Emlen made his fake firmament revolve around Betelgeuse, a bright star in the constellation Orion, the buntings flew as if Betelgeuse, not Polaris, were the North Star. By systematically removing and reinserting portions of his planetarium sky, Emlen found that his buntings relied especially on constellations close to Polaris, such as Ursa Major, Ursa Minor, Draco, Cepheus and Cassiopeia.

'These are the Flint Hills,' Jean said. 'In daylight this is beautiful open country, very green.'

She paused.

'The Flint Hills of Kansas,' she said, as if quoting the title of a plangent popular song. Another pause. I gazed out of the window at all the passing lights. The curling brunette wig lay unclaimed on the dashboard.

'We had a laundry in the convent,' Jean continued. 'A professional laundry. It was one of the ways we made money. We used to laugh because we knew we were washing Mother Superior's pantyhose or the chaplain of the University's boxer shorts. We were like naughty schoolgirls. We loved to giggle. We washed underthings, delicates, petticoats, scapulars, wimples, veils, habits, bedclothes, you name it, pressing and starching, finding a partner to fold the sheets with. We used rollers to squeeze out the water, and big industrial presses. We were supposed to keep silence. You just heard the machines going. I started to love laundry, the smell of clean clothes and the way they feel.'

'I know what you mean.'

'Do you? I *love* laundry. I love to do my husband's laundry. I love to wash clothes for friends of mine. I think of myself as a radical feminist, but I love doing people's laundry. For me, the best way I can show affection, or the warmth I feel towards someone, is to launder their clothes. Delicates I like to wash by hand with soap flakes. I've got a wicker basket for things waiting to be ironed. Don't you love the smell of fresh laundry? Sometimes when I tell my so-called feminist friends that I love to wash my husband's clothes, I hear them tut-tut as if I've committed a crime, but I don't think it's a crime that I like doing laundry.

'I don't have a drier. I hang things on a line in the garden. We visited Venice a few years ago. We walked down lots of these alleyways with beautiful old houses on either side and washing-lines strung between the houses. There was all this fresh laundry hanging over our heads – shirts, sheets, dresses, brassières, colours, and whites with the sun in them. All those bright colours. The shirts were waving like flags. When I walked under those clotheslines I felt like a bride walking under arches of fresh flowers.

'I had a friend who passed away last year. She'd been sick for four years. I went to see her pretty much every day and did all her laundry. I made sure she had clean clothes and clean bedlinen. I thought that if she had clean bedlinen that would make quite a difference to how she was feeling. There was one particular nightgown she liked to wear. It was very thin cotton with lace around the neck and on the shoulder straps. I washed that nightgown by hand over and over again, and she was wearing it when she died. I felt very close to her then, because we were such friends, and also because she was wearing the nightgown which I'd washed. I don't like to hear anyone say I'm wrong to love doing laundry.'

Jean reached down for the tote bag that lay between her feet. She lifted it to her lap, rummaged briefly, and pulled out a postcard.

'I like to collect things that have to do with laundry. I brought this along to show my sister. Maybe you can guess what it is.'

I looked at the postcard: a surreal, near-photographic painting of a washing-machine, not a commonplace household washing-machine but something like a large earthenware bowl, painted grey, with a chunky lid on top, and a round window in which a jumble of clothes was visible: it seemed antique and futuristic at the same time.

'No,' I said. 'I don't think I can.'

'It's Mickey Mouse's laundry room,' Jean said.

'Really?'

'Yes! It's from the Disney Museum.'

Yes, I could see one of Mickey's yellow mitts pressing against the window. Containers of laundry-related products were ranged along a shelf: a box of Freeze Detergent (For Really Cold Water), and bottles of Toonox Bleach, Toony Fabric Hardener and Toonite Liquid (For Fine Washing). A small barrel of clothes pegs hung from a rail. Each smooth peg had been carved from wood and resembled an elegant chess piece.

'Up on my kitchen wall I like to stick photos of my friends' laundry rooms. I have a needlepoint picture of women doing laundry. It's from a painting by Clementine Hunter. She came from a slave family on the Melrose Plantation in Louisiana. There are three women doing laundry, wearing dresses, orange, lemon yellow, the hottest pink you can imagine, and a big black kettle with a fire underneath it – one woman stirring the cauldron and the other two leaning over baskets, about to hang clothes on the line. I've got a collection of clothes pegs and laundry pins – old wooden pegs without springs, like these here' – she pointed to Mickey Mouse's clothes pegs – 'and all sorts of sprung plastic pins, every colour you can think of, transparent, opaque, milky, glittery. I don't need to tell you I'm proud as a peach of my laundry collection!'

I lifted myself up in my seat and looked back down the bus at people sleeping, the Greyhound a gallery in which diverse attitudes of repose were on display: heads tilted back, mouths agape, necks limp, cheeks on shoulders, couples slumped together, all lit up when the Americruiser cruised through concentrations of streetlights at the intersections, and all eyes closed but for those of the two white-bearded Amish elders, who looked straight back at me with the inscrutable, wild gaze of prophets. Tail-lights moved in the traffic flow like red-hot coals in lava streams, and sometimes the line of Interstate 35 appeared ahead of us, a light-course bending eastwards, not perceptibly founded on solid ground, but airborne, like the tube of bats that had curved away from Congress Avenue. I imagined this rope of lights as something useful to migrating birds, a guideline, and thought of flocks flying above us, town and city lights arranged beneath them in fixed constellations: zodiacs above and below.

The mechanisms of avian orientation are not fully understood. In species that migrate in flocks, including ducks and geese, experienced birds may guide juveniles from breeding grounds to winter grounds and back again. Birds are known to inherit an endogenous programme for migratory activity; to navigate using solar, magnetic and stellar compasses; and to pilot by familiar landmarks. It has also been suggested that they find their way by reference to winds, smells, infrasounds and minute changes in gravity and barometric pressure. 'Birds,' Emlen wrote, 'have access to many sources of directional information, and natural selection has favored the development of abilities to make use of them all.'

We came to Kansas City and waited in the terminal for our connection. The Greyhound for Minneapolis got under way after midnight, helmed by a younger driver, a man in his mid-thirties, spick and span the way a house can be, with a neat, trim moustache and a smooth, shining bald pate like a cap of polished wax, his uniform exemplary in crease and aspect, his announcements crisp and honed – he appeared just-minted, like a new coin. Jean and I sat together, two rows back on the right-hand side. The terminals received travellers and discharged them in fresh combinations: we recognized some of those who had boarded the coach with us, and noted the absence of others, like the Amish elders, who had boarded coaches assigned to other reaches of the network, bound for other destinations. Across the aisle sat a grey-haired man, jowled like a bull seal in a green suit, his tie loosened and top button undone, and before the Greyhound reversed away from the terminal gate he addressed himself to Jean and me, saying, 'I'm just waiting for the wheels to get turning. As long as the wheels are turning, I'm getting closer to home.'

Soon he was asleep. Jean slept. She had removed her glasses; there was a moist pink groove on the bridge of her nose. She slept with her head straight, tilted back on the headrest, mouth open, hands resting on the black tote bag across her lap. I slept, woke, and slept again as we continued north up Interstate 35, continuing north with the snow geese across Missouri and Iowa into Minnesota, between the Great Plains and the Great Lakes, sleeping when the coach hit cruising speed and the wheel-drone settled to an even pitch, lights spinning past in regular cadence, and waking whenever such constancy was interrupted, opening my eyes to find Jean asleep next to me, shoulder to shoulder, dreaming of tennis and fresh, fragrant laundry.

Laundry. We had a laundry room, with a drying rig of dowels raised by ropes and pulleys, so that you hoisted the wet sheets and towels like sails, and if you needed to walk from one side of the room to the other you'd have to part the drying clothes with outstretched hands as if they were lianas and fronds, or else give in to the clamminess of damp shirts and trouser legs as they dragged across your scalp and cheeks. There was an old wringer with crank-turned rollers, and an oversized paint-stained sink beneath a shelf that was crowded with bottles of Brasso, bleach, turpentine, household ammonia and limescale remover, and also with paintbrushes, scrubbing brushes, yellow Johnson's wax polish and beeswax polish and a rusting pink Flit fly-gun with a trademark white-trousered soldier marching on its canister. This white-trousered soldier could himself be seen toting a pink Flit fly-gun, and once, when I was very young, I studied the soldier's fly-gun to see if I could find on its tiny canister an even tinier soldier toting a fly-gun, imagining an infinite, shrinking series of quantum soldiers toting fly guns. And opposite the washing-machine, on its own square concrete plinth, rested the old blue oil-fired boiler, the house's heart, with a complex system of padded arterial white pipes leading from it to the ceiling. The small room in which I'd slept as a child (and in which I slept when illness returned me to the condition of a child, dependent on my parents, unable to cope with the challenges of the world outside my immediate home range) was directly above the boiler, and each morning, soon after the rooks began cawing, I'd hear it shudder to life as the timer decreed – the walls shaking, the table-flap rattling on its secret latch, a sound in the floorspaces as if big, clumsy bubbles were galumphing up the white pipes, carrying heat to the house's extremities.

It was not hard, sitting on the Greyhound that night in March, following the snow geese, with Jean asleep beside me and the mesmerizing, hallucinatory flare and slide of lights all around us, to return home, to go back to the laundry room, or to the short white passage that led from it to the back door of the house, where the bars of three bolts slid with known weight and easiness into sockets on the jamb. The door opened on to the small paved terrace, the feeder with its red-husked peanuts, and if you looked to the left you'd see chestnuts, sycamores and limes, you'd hear the bassoon caws of rooks in the tree crowns, the sound of the Sor Brook dropping off the waterfall, and if you looked to the right you'd see shrubs and climbing roses along a wall, a copper beech, farmland receding in a gentle upward grade to the west, the fixed pattern of fields named Lower Quarters, Danvers Meadow, Morby's Close, Allowance Ground.

Illness had taken me back, the first time since I was a schoolboy that I'd spent more than a few days, a week at most, at home. And it *was* home: the fact that I hadn't lived there for years didn't change that. Nowhere was my sense of belonging so unambiguous. I could still find my way around the ironstone house in the dark, or with my eyes closed, moving by reflex, habit, muscle memory, my hands knowing just where to reach for a handle, switch or rail, my feet ready for a step up or down, a loose board, a shift from carpet to stone. I knew the names of things, their details, histories, every surface burnished with memory and association. When I fell ill, feeling threatened, under attack, with all sense of control or mastery gone, I longed for the house, imagining a place of safety, without dangers or conflict, where all my needs would be provided for, a still point from which life's unsteadiness could be viewed and measured. But the longing was a fantasy of escape. It was nostalgia.

I opened my eyes. A grey morning, dull, pewter-toned, twenty-four hours since I'd left Austin. The Greyhound was cruising steadily. Jean was awake, rubbing her eyes, replacing her glasses.

'Did you sleep?' she asked.

'Sort of.'

'I was out for the count. Oh my, I was bushwhacked.'

And without warning she flung her right arm across me, pointing eastwards.

'Look!' she said. 'Those are geese, right?'

Yes, those were geese. Flocks of snow geese were flying in skeins and straggling Us of thirty or forty birds each, moving northwards over flat country, above the horizon, parallel to the Greyhound. Slow waves rode through the strands as leading birds deviated slightly from their straight course and birds behind them followed suit, one after another, passing the discrepancy like a rumour along the line until it reached the last bird and flicked out into open air. Even in the grey light I could distinguish blue-phase from white-phase birds – the morphs not jumbled randomly along each skein but grouped together in bands of three or four geese of the same colour. Each group of blue-phase or white-phase geese probably represented a family, two mates and their young: snow geese pair for life and forge strong family bonds, parents and offspring staying together on the first migration south, during the winter, and on the spring migration back to the breeding grounds, with the male usually leading his mate in flight – the opposite of ducks, where the female takes the lead. Jean leaned over me, getting her face close to the window, craning for a better look at the snow geese, and for a few minutes we kept level with the flocks, until the Greyhound pulled ahead, bent on Minneapolis. I was wide awake now, heading north with snow geese, complicit with birds.

We arrived shortly after nine o'clock. Jean was anxious to see her sister. She got into the back seat of a taxi, looking surprisingly fresh, ablaze with red, yellow and blue, the black tote bag hanging from one shoulder, the badge still adamant on her fleece vest. She wound down the window.

'I hope things are all right,' I said.

'I know. We'll see. I hope those geese haven't skipped town when you get there.'

'Bye, Jean.'

'Bye!' She was calling it out; the taxi was already moving.

There was a bird in the terminal at Minneapolis, a passerine, perhaps one of the sparrows, skittish in the roof girders. Exhausted, disorientated, porous with distance, I sat watching a tall, swarthy man keep a leather purse of beans up in the air, the little sack jumping from knee to knee, the arch or instep of one foot to the arch or instep of the other – a hypnotic, stringless yoyoing accompanied by rhythmic percussive beats as beans scrunched together in the impacts on leg or shoe. I boarded my last Greyhound at noon, limbs aching, the coach proceeding north-west on Interstate 94 towards Fargo, crossing from Minnesota into North Dakota. Farmed country ran flat in all directions, as if the land had conceded to the sky's magnitude, and given ground. Intricate centre-pivot irrigation systems stood motionless in the ploughed black fields. Farm buildings hunkered down in horseshoe windbreaks. I looked for geese, black wing-tips flickering in smoky white sky. The coach passed St Cloud, Sauk Centre, Alexandria and Fergus Falls, and it was close to six o'clock when we arrived at Fargo. The temperature had fallen dramatically. Thirty-five hours had passed since the Americruiser had pulled out of Austin in the storm. Eleanor's shining birdcage stood empty 1,000 miles due south as the crow flies.

4: SAND LAKE

GIDEONS HAD LEFT A BIBLE on the bedside table of my motel room. The bible's brown cover was embossed with the Gideon logo, a flame tufting like Tintin's hair from the mouth of a gold amphora. In its introductory pages, before even Genesis began, one verse from St John's Gospel ('For God so loved the world, that He gave His only begotten Son, that whosoever believeth in him should not perish, but have everlasting life') was translated into twenty-seven languages, and if you followed the word 'son' as it mutated through these tongues – *son, søn, zoon, Sohn, sinn, sønn, seun* – it was as if you were looking at an intimate's face in different lights and moods. No pictures hung on the white walls. An enormous black television presided over the square room; a receiver dish hummed with frequencies outside the sliding windows. Between the bed and the windows, facing the television, was an old La-Z-Boy reclining chair upholstered in brown corduroy, its arms welted with cigarette burns. Glasses stood rims-down on a red tray. A band of paper girdled the toilet seat like a cummerbund to emphasize the motel's attention to hygiene.

From Fargo I'd driven a rented Mercury Topaz south and then west to Aberdeen, South Dakota. Midcontinent snow geese tend to set down on the lakes of North and South Dakota to rest and feed while the thaw advances northwards ahead of them, and the fact that Jean and I had seen flocks flying hard towards Minneapolis suggested that geese were already arriving in these latitudes. I drove in dense white fog, glimpsing fence-posts, barbed-wire reels, grain elevators, stands of cottonwoods, and prairies rolling away in mottled browns and greys. My expectations quickened whenever the fog thinned. I looked from side to side, or leaned forward over the wheel for a fuller view of the sky, hoping for the flicker of a black-tipped wing. The highway crossed frozen lakes on low causeways. Strong southerly winds carried snow and spiculae off the surface of the lakes in sheets that dragged across the asphalt or swept sudden granular hisses across the windscreen of the Topaz. The diesel stacks and radiator grilles of freight trucks loomed in my mirrors. Road signs emerged from the fog like things I was remembering. Many bore the names of English towns: Bath, Bristol, Andover, Stratford. Migrants had travelled with proper nouns as though they were personal effects. The names were tokens of home.

*

SNOW GEESE ARRIVE! The next morning, the front page of the *Aberdeen American News* reported that more than 340,000 snow geese had arrived at Sand Lake National Wildlife Refuge in the last twenty-four to forty-eight hours. I couldn't believe it: I'd reached South Dakota on the same day as the geese. Sand Lake was not quite thirty miles from Aberdeen, due north across the Dakota Plain on a quiet, empty road through flat country, with open, sere-looking grasslands and close-cropped stubble fields on both sides, and the roundels of small frozen pothole lakes like a strew of blue-white doubloons across the prairie: no greens, just tans, fawns, duns, greys and the blue-blotched white of the ice. American kestrels, the smallest of the North American falcons, perched on telegraph wires, compact and snug, tucked up in the bed of their own grey and rust-red feathers. Flocks of gleaning red-winged blackbirds rolled across stubble fields like giant, shadowy tumbleweeds. I'd never paid so much attention to birds. I kept a field guide and my birdwatcher's binoculars within arm's reach on the front passenger seat. I looked out for snow geese. I couldn't stop thinking of snow geese. Three hundred and forty thousand snow geese.

Sand Lake was a long swell in the James River, close to the border with North Dakota. I parked the Topaz next to two white pickups with the US Fish and Wildlife Service shield on their doors. I buttoned up my down-filled jacket and hung my binoculars round my neck. It was very cold. There were stands of bare cottonwoods and elms, shelterbelts of green ash and Russian olive, shrubby willows in the marshy ground. The sky was deep blue, beyond dimension, appearing to curve away with the curve of the Earth at each extremity. Clouds resembling judges' wigs drifted on the blue. I could hear, in the distance, a faint, familiar sound, a great crowd of terriers yapping at the limit of earshot.

Excited, I began walking north along the track of dry dust and stones that ran the few miles from the refuge entrance to Houghton Dam. Thickets of cattail rushes and phragmites made a golden rind around the lake, stems clamped in ice at the shins or ankles, the cattail tipped with stiff brown seedheads like fat cigars. Sometimes pickups driven by hunters wearing camouflage cruised past me on the dirt track, each vehicle's slipstream agitating the cattail and phrags. The yapping thickened to a drone. I passed a small farm, then rounded a headland, walking faster and faster towards the source of the noise. Snow geese came into view like a kept promise. Thousands of blue-phase and white-phase birds were huddled on the ice in the middle of the lake, a huge white almond-shaped spread tapering to a point at its north and south ends. The birds' heads were raised high, their necks extended, perpendicular to the ice. Close to, the flock's gabble was a wild encompassing din, the birds' calls travelling through the ice like marbles rolling on metal. I stood still, breathing deeply, half-hidden by cattail.

A Fish and Wildlife Service pickup pulled up beside me, and

a man in the neat buff uniform of the wardens leaned from the window. He was about fifty; he wore steel-framed glasses; he had thick, emphatic eyebrows and pellucid grey-green eyes. His trim hair was greying: a mixed flock of blue-phase and white-phase strands.

'Not bad, is it?' he said.

'It's the biggest flock I've seen. How many?'

'I'd say thirty thousand. They're real early this year. Usually you can set your clock by snow geese getting to this latitude. It's always the last week in March. They've taken us by surprise. We thought, maybe another ten days, a week at least.'

'Why are they so early?'

'Spring's early. They come with the weather. These geese push up as far as they can. If they run into storms or it's too cold for their liking, they just rest up. Some of them go north to take a look and come straight back if it's too rough up there. I'm Michael, by the way. I've got a feeling you're not from South Dakota.'

Michael switched off the engine, and joined me on the rise overlooking the lake. He was tall, over six foot, his black boots buffed like a soldier's, his belt sagging with professional things: revolver, cuffs, pepper spray, retractable baton. We looked at the geese, talking against the background of their din.

'You have an interest in geese?' Michael asked.

'Snow geese. I've just come from Texas. I'm trying to follow snow geese from their winter grounds to breeding grounds. Just keep with them on the spring migration.'

'You are? That's something. How far north are you going?'

'Hudson Bay. Maybe Baffin Island.'

'That *is* something.'

'It's not exactly a sensible way to travel. I'm at the mercy of geese. I'm starting to wonder if that's such a good thing to be.'

'I've spent pretty much my whole life chasing ducks and geese one way or another, and don't think I haven't asked myself the same question.'

'Did it surprise you, that they got here so early?'

'I guess so. But patterns are always changing. I mean, look at the eagles. Before I came to Sand Lake they hadn't had a documented eagle nest in South Dakota since 1885, which was four years before South Dakota even joined the United States. In 1991 a pair of bald eagles showed up and nested in Sand Lake Refuge. We cordoned off the area with a half-mile buffer zone. The eagles were incubating – if you looked through a scope you could see the male and female changing places. But this windstorm came through and frazzled the nest. The eagles folded their tent, and then these damned great blue herons moved in and threw an egg out, which we found underneath the tree. In 1992 another pair came and nested on Karl Mundt Refuge and raised one young. That pair has been successful every year. Now there are four or five nests in the Sand Lake vicinity. The population's exploding. You've got bald eagles nesting pretty much everywhere.'

I wondered if that had something to do with the supply of snow geese.

'Sure. The eagles stick with the geese. They go for sick ones. Crippled geese gather in the last spots of open water. I've seen them diving to escape an eagle, and an eagle actually going in the water, like an osprey would, to grab them in the talons and drag them up on the ice. They can't fly off because they're waterlogged in the feathers. They shake themselves off on the ice, holding tight to the goose that's flailing around, and preen themselves a little bit, and then get on with the business of eating goose. They've got fish, too. Eagles and herring gulls come up early for the fishkill. This ice is two feet thick, but a dead fish underneath it is a dark spot that absorbs more heat and melts its way up to the surface, and that's where the gulls and eagles'll have it – '

He stopped: there was a commotion in the flock. The calls of the geese grew louder, more urgent. Suddenly, as if detonated, the flock took wing. Thirty thousand geese lifted off the ice in front of us, wingbeats drumming the air, goose yelps gathering to a pounding, metallic yammer, the sound of steel being hammered on anvils, in caverns. The ice thrummed and sang with it. The exploded flock filled our fields of vision, a blizzard of birds. Most of the geese flew in low circles, but some settled back on the ice almost immediately, while others continued to gain height. Drifts of geese passed through, behind and across other drifts of geese; the flock kept wheeling round and round, swirling with eddies and countermotions, a salt-and-pepper chaos of blue-phase and white-phase birds lit by quick lamé sparklings of white wingbacks catching the sunlight. Whole swatches of the flock went dark when birds flew side-on, and swatches flashed white when they banked or veered, breasting the light. Then slowly, goose by goose, the flock settled again: the almond shape reformed; the extravagant din dwindled; the steady flock drone resumed. For a moment, I had forgotten to breathe.

'Look at that,' Michael said, pointing. 'That's the reason those geese are so uppity.'

On the far side of the lake, a few yards in from the cattail, an eagle stood on the ice, steady as an urn, with a blackish body and distinguished white head: a bald eagle, looking for a sick or injured goose. The snow geese were anxious, vigilant, ready to fly. Small troupes were taking off from the flock and circling; others were coasting down, wings bowed, landing in the almond: an ongoing exchange and renewal of component cells. The eagle kept its vigil close to the cattail. Cloud shadows drifted on the ice.

'I'll see you again,' Michael said. He got back into the pickup. Dust billowed from the heels of the truck as it pulled away. I walked all afternoon, right around Sand Lake on the dirt track.

*

EACH MORNING I drove out to the refuge to watch birds. I took lunch and stayed all day, walking the track to Mud Lake, crossing the James River at Houghton Dam, watching white-tailed deer venture from windbreaks of elms, the sun going down somewhere on the far side of the Missouri River, pronouncing 'West!' as it sank in the grasslands and buttes. Deer bounded along the stubble edge as geese returned to their roosts from grain fields: smudges and specks of geese above the low sun. Geese flew from the south in long skeins and echelons that crossed and undulated, or appeared, by a trick of angle and distance, to twirl in ropes and double helixes. Geese flew in their limited alphabet of Vs, Js and Ws, or in interlocking chevrons like the insignia on officers' epaulettes, and the high raillery of snow geese in flight made a descant to the deeper, rougher honking of Canada geese roosting in the phragmites and cattail.

More and more snow geese arrived at Sand Lake in the last days of March. Gulls appeared, scouting for fishkill; ducks settled on patches of open water by the dams. Seasoned birdwatchers, wearing sweatshirts in pastel colours with sandpipers or whooping cranes printed over the heart, drove out from Fargo and Minneapolis, parked Dodge and Mercury minivans and Jeep Cherokees on the shoulders of the track, and lugged Cullmann and Manfrotto tripods and Optolyth, Swarovski and Questar telescopes in padded green sleeves to vantage points on the lakeside. The appearance of the birdwatchers was itself a kind of migratory return.

Michael invited me to join him on his bird counts. His uniform was neatly creased and spotless; his black boots were unimpeachable; his belt, with its ballast in black leather pouches, sagged low on his hips. Often he took off his steel-framed glasses with his left hand and rubbed both eyes with his right. Before replacing the glasses he would blink dramatically, as if he'd just been swimming underwater.

Michael had trained as a limnologist: he knew the ins and outs of ponds. In his twenties, when he first came to Sand Lake, he'd made a floating hide by fixing plastic pipes in a domed frame, covering the frame with burlap bags, camouflaging the burlap with cattail and phrags (thatching the rushes so the hide resembled a muskrat house) and mounting this canopy on an inner tube from a tractor tyre. Wearing waders, hidden beneath the dome, he could move through rushes without scaring birds. He remembered the first time cattle egrets nested at Sand Lake, in 1977, and he liked to think of the egret gradually extending its range, crossing the Atlantic from Africa to South America in the nineteenth century, spreading north over the Caribbean to Florida, and working up through the United States until it was spotted in South Dakota, in 1961, at Sand Lake.

We drove through the refuge in Michael's white pickup. I had binoculars; Michael had an old Questar scope that fitted to his wound-down window. A label warned users not to look directly at the sun, making you think of daylight concentrated to a laser in the tube. Michael pointed out corn, wheat, barley and soybean stubble; switchgrass, bromegrass, wormwood sage and little red shoots of smartweed; nesting boxes for wood duck, kestrels and bluebirds; muskrat houses in open ponds. He said that sometimes Canada geese built their nests on top of muskrat houses: homes on homes. Richardson's ground squirrels, which Michael called 'flickertails', slipped down holes in the tracks as we approached. A dog ran out from a farm, snapping at the pickup, and Michael laughed, saying, 'There'll be toothmarks on the fender!' Whenever we came across a flock of snow geese on some part of the lake, he stopped the truck and fitted the telescope to the glass. He wrote down figures on a pad: 28,000; 16,000; 45,000. The flocks were growing. White-phase and blue-phase birds stood close-packed in islands on the ice.

Michael taught me to identify ducks. The males of each species were easier to distinguish: lesser scaup with their black heads and white bodies; the similar ring-necked duck with the white ring at the base of their bills; redhead and canvasback with their reddish heads and pale grey backs and sides (canvasback distinguished by their flatter, sloping foreheads); goldeneye with their golden eyes and the round white spot on their faces; bufflehead, nicknamed 'butterballs', with their white headdresses; the long slender necks of northern pintail; the green heads, white bodies and shovel-like spatulate bills of northern shoveller; the flash of livid blue on the wings of blue-winged teal.

Learning the names is a method of noticing.

'That pair next to the ice,' Michael quizzed me. 'What are they?'

'Scaup?'

'No!'

'Goldeneye?'

'No!'

'Ring-necked?'

'Ring-necked!'

It seemed that what a thing was called was a part of what it looked like. The ducks were transformed (fleshed out, coloured in) when I matched them to names: bufflehead, wigeon, gadwall. We watched seven male canvasback compete for the attention of a single female, raising their bills, lunging like fencers, and male goldeneye throwing their heads back in courtship displays, as if swallowing aspirins. Michael could tell species from their styles of flight (the twisting, turning flight of compact teal flocks; the low, single-file processions of common merganser, diving ducks with white bodies and slim red bills) and knew in just what sequence to expect them back each spring.

'First up are Canada geese,' he said. 'Then the mallard and snow geese, bald eagles with them, and then a lot of the diving ducks, goldeneyes and mergansers, which are fish-eaters, not much good as table birds, and now we get into redheads, canvasbacks, red-tailed hawks are here too, and sparrowhawks, and ruddy ducks and teal, blue-winged first and green-winged later, and now the water's opening up you'll get the white pelicans, double-crested cormorants, great blue herons, cattle egrets, black-crowned night herons, killdeer, Franklin's gulls, Forster's terns and black terns, sandpipers, western phalaropes, pied-billed grebes and eared grebes and western grebes and then the songbirds start . . .'

He paused, removed his glasses with his left hand, rubbed both eyes with his right, and blinked three times with considerable purpose.

*

MICHAEL INTRODUCED ME to Rollin. He was eighty-two years old, with soft white hair and a long white beard: he had the white head of a bald eagle. Like Michael, he wore steel-framed glasses, but Rollin's glasses had thick lenses in which his eyes floated like puffer fish behind aquarium glass. The glasses sat with a look of stolid permanence on his robust, beaky nose. He was fit; he had the energy of a man twenty years his junior. He wore hiking boots and the kind of multi-pocketed sleeveless vest favoured by anglers and professional photographers. He had driven from Iowa to see the eagles at Sand Lake.

'Mike says you like to walk,' said Rollin.

'He's right.'

'He thought you might be interested in taking a walk. There's a good number of eagles in the cottonwoods out past Houghton Dam. I saw them on my way in.'

We started off up the track alongside the lake. We both had binoculars hanging round our necks, the binoculars knocking rhythmically against our chests as we walked, beating on our hearts like pacemakers. We walked with a crop of maize to our left, cattail and phragmites to our right. Beyond the rushes, on the far side of the lake, we could see a flock of snow geese, at least 30,000 birds. The calls of the geese telegraphed through the ice as if people were banging on it with hammers.

'What got you interested in birds?' I asked Rollin.

'I've been lucky, I guess,' he said. 'I grew up in north-west Iowa, east of Sioux City about sixty miles. I did my schooling in Cherokee County, and between my junior and senior year in high school I went up to North Dakota for a year to live with my uncle and aunt. They were farmers. Raised wheat, flax, rye, oats. They were losing money. Drought was killing all the crops and it was mighty tough times in the Depression. It was in North Dakota I found my love for the out-of-doors. My parents were interested in wildlife, but no more than anyone else. My uncle knew a lot of the names and such, and I looked up a lot in books. I don't know. It was me, see? It was something in me. I just had to know. I always had an appetite for knowledge. I just kept asking questions. In North Dakota I just fell in love with marshes and sloughs. I saw my first eagle in North Dakota. Caught my first trout in North Dakota. Shot my first duck in North Dakota. It was a spoonbill, of all things, but I didn't know the difference. I was using a double-barrelled 12-gauge and like a dummy I pulled both triggers at the same time and got sat on my ass real fast. I hit the duck and found it later, but I didn't think I'd hit it because the first thing I know I'm getting up out of the water. Got kicked over by the recoil, see? I wasn't very big at the time.

'When I was in high school I lived right by the Little Sioux River, and the Little Sioux's a beautiful north-west Iowa stream. I went up and down that river an awful lot, fishing and camping. In high school I didn't play football or basketball or baseball. I was uncoordinated and nothing worked right for me. But I always liked to shoot, see? I got to shooting on a big-bore rifle team, an adult team when I was in high school, and I got to be pretty good at it. I was an expert rifleman in .30 calibre, and in .45, and .22. I could do that, see? A lot of sports I couldn't do because I had to wear glasses, but I liked to shoot, and that's part of how I got to be an outdoorsman.

'After the war I moved back to Iowa and got out-of-doors every minute that I could. I liked books like *Sand County Almanac* by Aldo Leopold. I liked Zane Grey westerns. Do you know *The Vanishing American*? *Man of the Forest*? *Lost Pueblo*? Do you know *The Light of Western Stars*? I had a job with the postal service. I was a real letter-carrier for twenty-seven years. Sorted and delivered mail to three hundred and eighty families on a hundred-mile route, most of them farmers. Soon as the route was done I'd get outside. I classify myself as a short grass prairie kind of guy. I love the prairie. I have a philosophy that every one of us is on this earth for a reason. I try to share my love of the beauties of wildlife, how wildlife is part of us, because they demand the same things we do, and if they can't make it, we sure as hell can't either. I'm partial to birds. Eagles, hawks. Songbirds I like real well, but eagles I like best of all.'

'Why eagles?' I asked. We were walking side by side along the track at a good, even pace, binoculars knocking on our chests. Rollin walked with the rolling, side-to-side motion of a galleon, and his voice had the cheery, tuneful fluency of a sea shanty sung on board.

'Eagles,' Rollin said, 'well, eagles are top of the line. King of the hill. Their eyesight is fantastic. They reckon their eyesight is two and a half times better than yours and mine. There's a tremendous pecking order among eagles. You've got an immature bald eagle sitting on a tree branch, and along comes a mature bird who wants that perch – it'll just bump it right off. Or if an immature's eating a fish and a mature bird wants it, it'll come straight in and knock it off, and there won't be an argument about it. A bald eagle will usually perch on the highest branch it can, and if they find a cottonwood tree all the better. They want to look around. They like to know what's going on. A hawk will quite often take a lower branch. An eagle isn't afraid of any other bird. A lot of birds have to perch according to what they think is their fear level of being attacked, but a bald eagle has no fear of other birds.

'Whenever I get the chance to see bald eagles, I take it. One winter I saw bald eagles on the Mississippi River, near Davenport, below the locks and dams. The river was frozen right up to the spillway. There wasn't much open water. The eagles picked off fish that went through the lock, through the rollers. Fish got kind of stunned by the rollers, and eagles grabbed them while their heads were swimming. I've seen eagles in Manitoba, Missouri, Louisiana, Yukon Territory, Alaska. There's a two-mile stretch of the Chilkat River in Alaska where it's not unusual to see two and a half to three thousand eagles, drawn by the late salmon run. I've sat underneath trees where bald eagles were so unconcerned that I was there, they were preening themselves, not paying any attention to me.'

We stopped. We could hear ducks chortling on the far side of the rushes. We raised our binoculars.

'Scaup,' I said, fancying myself an expert. 'A couple of pairs.'

Rollin raised no objection. A Canada goose honked in the cattail.

'Those Canadas,' said Rollin. 'Those are proud birds. I like them.'

We started walking again, heading on towards Houghton Dam.

'What happened when you finished high school?' I asked.

'Well this was the height of the Depression,' said Rollin. 'There were no jobs. You did what you could. You did whatever you could find. Maybe you worked on a farm a few days, and you worked on a threshing crew. You just did whatever came. I worked in a grocery store, but I knew that couldn't last because I wanted outdoors work. I sold advertising for the local paper. Then the war started up, and I entered the Navy as an aviation gunnery instructor. I guess they saw my background in newspaper work, because I got to work for two different station newspapers. One of them was at Monterey, California, and one was at Santa Rosa.

'While I was at Santa Rosa we had a guy come from the South Pacific, a commander, a public relations man. He got word out to all the station papers that he would grant a free-wheeling interview to anyone who came up to San Francisco. So the chaplain, who was my boss, said, "Do you want to go, Rollin?" and I said, "Yeah, I'd like to go." They flew me up to San Francisco in a torpedo bomber, and we landed at Alameda Naval Air Station. I went across the bay, and up to the top of the Mark Hopkins Hotel, and I was in on the interview. Only thing is, this public relations man turned out to be an asshole if ever there was one. He was a full commander, a bigshot from Minnesota, and he kept us standing in this anteroom for an hour while he made small talk with his orderly. Finally he came in. He never told us, "At ease." The orderly had passed out sheets of paper with questions on them. "These are the questions," he said, "which you may ask the commander." Jesus! Oh, I never said a word. I never asked a question. There wasn't anything there but what you couldn't have taken out of the *San Francisco Examiner* that morning.

'Well, I went back across the bay. It came time to fly back to Santa Rosa. The pilot said, "Do you want a little fun?" and I said, "Sure. What have you got in mind?" So he said, "We'll fly under the Golden Gate Bridge!" Now, that was unknown of. If we got caught at it we'd get court-martialled for sure. Anyway, we decided to do it. We got back in the torpedo. I was in the gunner's seat, behind the pilot. The plane was camouflage-painted. They'd have had to catch us dead on, because we flew close to the shore, and Christ, a little ways away you couldn't even see us, let alone hear us. Joe, the pilot was, Joe something. I guess he'd always wanted to do it, and this was an ideal time to do it. There were very few planes in the air and it just looked all at once like this was the time we were going to do it. We took off and flew over the water towards the bridge. There's two hundred and fifty feet of clearance. As I recall, there was a boat coming at us under the bridge, but not a battleship, and I doubt like hell they saw us, because we just blended in with the background, see? The bridge isn't golden, it's a kind of rusty, orangey red. Over to the left there were hills and on the right there were beaches, dunes, big surf. Then you've got the Presidio, and there were houses there, but you don't look at the houses, you look at the ocean, and the city. So we flew closer and closer. And this Joe liked to say "Son of a gun!" all the time, and when we got to flying under the bridge he was screaming out "Son of a gun! Son of a gun!" like there was no tomorrow. I was just laughing. The ocean was so close you could hang your hand out the cockpit and skim it just like a bird. So we flew under, and that was kind of fun, see?'

We were almost at Houghton Dam.

'Last summer,' Rollin continued, 'I was in California, visiting relatives. My sister-in-law has a friend who works in tourism in San Francisco. I called her up and said, "Is there any way we could get to the top of the Golden Gate Bridge?" She said, "You know, that's almost impossible, but I'll see what can be done." Then I didn't hear a thing. And then a few days later she calls and says, "Call this number." So I did, and it was the Transit Authority himself. I told him why I wanted to go to the top of the bridge, on account of my having gone underneath it illegally fifty years ago, and also I'd been to the top of the Straits of Mackinac Bridge, where Lake Michigan goes into Lake Huron. He thought a good long time and then said, "Well, normally I'd say no, but you've got a good story. We'll let you go." I went with a friend of mine. He'd been in the navy too, and he'd lost his wife a few months before. We drove down to San Francisco in the pickup, and we got to the top of the Golden Gate Bridge. What a view we had! God, that was fantastic. That was something. When I got to the top, the bridge was *swaying*. That's the way they're built, to do that in the wind. And this bridge just talks all the time. It *moans* in the wind, and the wind blows a gale at the top. You have to hang on. You can't get blown off, because the railings are high. You can see through the rigging, seven hundred and fifty feet down to the water. That was something. I won't forget that.'

We reached the dam at the north end of Sand Lake, and turned east towards Houghton. We walked without saying much for a while, following the road to the floodplain. A narrow causeway carried the road over the ice, floodwater frozen to left and right. Dead cottonwoods stood knee-deep in the ice, calcium-white where the bark had fallen away.

'Look,' Rollin said, stopping me with his right arm.

Three bald eagles were standing on the ice. They stood absolutely still. We watched them through binoculars. I saw, close up, the white head, blackish body, heavy yellow bill; the yellow feet and ankles; the severe eyes.

'Not those,' Rollin whispered. 'Look behind them.'

I looked. Between fifteen and twenty bald eagles were perching in a dead cottonwood at the edge of the floodwater lake. The tree resembled a candelabra, the eagles' white heads like flames on thick black candles. One eagle took flight from a limb of the tree. The branch juddered. The eagle fanned its white tail as it banked away from the cottonwood, gaining height. Further away, beyond the cottonwood, more eagles were gliding in circles, lifted on a thermal column: a kettle of twenty or thirty birds soaring without effort or fluster, in carousels, turning and turning on the updraft.

'Isn't that something?' Rollin said quietly. 'Aren't those the grandest birds?'

*

THE CONTENTS OF MY BAGS began to colonize the white motel room. Bird books and papers piled up on the desk, among them a copy of *The Snow Goose*, with my grandmother's initials pencilled inside the front cover, alongside a date: October, 1942. My father had found it among his books and left it on my bed in the dressing-room towards the end of my long stay at home. The red-tailed hawk's feather that Eleanor had handed to me beneath the radio mast lay flat like a bookmark between the covers of *The Snow Goose*. My possessions were beginning to breach the anonymity of the white room. Books and papers on the desk, my clothes on the floor, the room becoming familiar in spite of its blankness. Standing in the corridor, holding my room key, I knew what to expect when I opened the door: the black television, the La-Z-Boy, the two glasses standing rims-down on the red tray. I knew where to feel for a switch on the neck of the bedside lamp, as a nurse knows where to feel for a pulse. I could find my way around the room in the dark.

Each morning I drove the Topaz north to Sand Lake. I walked and watched birds, sometimes with Michael or Rollin for company. But at night, in the motel, I became anxious. I'd been away for a month now, chasing geese. There were people I missed. I felt the allure of familiar scenes, the pendulum of my impulses swinging back again, swinging back from what was new and undiscovered towards all that was known, named, remembered, understood. I was lonely in the white room. For the first time I felt frustrated by the journey to which I'd committed myself. On maps, the flight of snow geese from the Gulf of Mexico to Hudson Bay and Baffin Island was a flawless, unbroken arc, the curve of time from one season to another. But the reality was different – not a smooth, continuous passage from here to there, but a stop-start, stage-by-stage edging towards the north, with geese flying from one resting area to the next, proceeding only as far as the weather would allow. I had attached myself to the birds. I couldn't move on until the birds moved on, and the birds couldn't move on without the spring.

One night, I took the Gideon bible from the bedside table and leaned back in the La-Z-Boy. Browsing aimlessly, I found that time after time the bible fell open at the same place, in *Psalms*. I looked closely. Someone had torn out a page. Page 617–618 was missing. Psalm 23 was missing. I wondered who had torn the page from the Gideon bible. The most famous of the psalms: 'The Lord is my shepherd: therefore can I lack nothing.' What in those verses had been so important to a previous occupant of this square white room? Or had someone, when using the telephone, simply reached out for any piece of paper on which he or she could scrawl a message? The missing page struck me with particular force, because only a few months before, I had copied the last verse of the psalm into a notebook: 'But thy loving-kindness and mercy shall follow me all the days of my life: and I will dwell in the house of the Lord for ever.' It had occurred to me how often the authors of scripture depict God as a house or shelter in which one might dwell, as if faith were itself a home, affording all the protection, comfort, steadiness and sense of belonging that home implies – as if the need for God were homesickness in paraphrase.

Homesickness. I'd never thought myself prone to homesickness. Even when I was eight years old and first going to boarding school, I hadn't been especially homesick. There was too much of the *new* to keep you distracted, to keep the mind off the charms and comforts of home. But I can remember how excitement mounted in the last days of term, how the mood picked up, our senses whetted by the imminence of return, so that when Mr Faulkner read us *The Snow Goose* (his long legs crossed at the ankles, the school's large white clock just visible through the high windows, like a full moon) you could feel a charged, febrile restlessness in the classroom, a quickening like static electricity. It was invigorating, the prospect of going home; it actually gave you energy. We whispered, fidgeted, passed notes, grated our chair legs on the wooden floor. But Mr Faulkner lost himself in the Great Marsh. He didn't seem to notice when the electric bell rang the end of a period. He kept on reading, persisting to the end of one paragraph, then another. He went on reading about Rhayader, the abandoned lighthouse, the retreat from Dunkirk and the migrations of geese, until the hubbub and clatter forced him to concede, and he marked his place with a tasselled bookmark, stowed the book in his ragged leather briefcase, and told us, wearily, we could go.

We could go. With this, Mr Faulkner acknowledged our restlessness and signalled release. But that restlessness was nothing compared to the longing I experienced when I fell ill. In hospital, after Christmas, the desire to go home was more powerful even than the desire to be well. Sometimes it was hard to distinguish one desire from the other. I waited for the doctors to tell me I could go home. Whenever doctors came into the ward, even doctors I'd never seen before (*anyone* in a white coat, with the pincer earpieces of a stethoscope protruding from a front pocket), my hopes lifted. I thought, *You are going to tell me I can go home.* And when, finally, my mother and father drove me back to the ironstone house, after dark, with a cushion tucked in under the seatbelt strap to guard the wound in my abdomen, I felt the grip of my anxieties loosen, I felt calm and lightened, as if I'd just been handed a reprieve.

In 1688, observing that homesickness lacked a medical designation, a physician from Mühlhausen named Johannes Hofer proposed that it should be known by the term *nostalgia*, which he had derived from the Greek words *nostos*, meaning 'return', and *algos*, meaning 'suffering' – 'so that thus far it is possible from the force of the sound *Nostalgia* to define the sad mood originating from the desire to return to one's native land'. For Hofer, homesickness was a serious disease whose symptoms were 'continued sadness, meditation only on the Fatherland, disturbed sleep either wakeful or continuous, decrease of strength, hunger, thirst, senses diminished, and cares or even palpitations of the heart, frequent sighs, also stupidity of the mind – attending to nothing hardly, other than an idea of the Fatherland.'

As an illustration, Hofer described the case of 'a certain country girl' who was taken to hospital after a fall. 'She lay prostrate,' he reported, 'without consciousness or movement for several days.' When the girl came to, finding herself 'handled about among the wrangling and querulous old women', she promptly fell victim to homesickness. She refused to eat, spitting out food and medicines. 'Especially,' wrote Hofer, 'she wailed frequently, groaning nothing else than "*Ich will heim; Ich will heim*," nor responding anything else to questions other than this same "*Ich will heim*." Finally, therefore, her parents allowed that she be brought home, terribly weak, where within a few days she got wholly well, entirely without the aid of medicine.'

Hofer saw that homesickness was an individual's response to a double challenge, a reaction not just to the loss of things you loved or took for granted in your old environment, but also to the strangeness of things you encountered in the new – 'the changed manners of living', the foreign climate, the food, 'and various other troublesome accidents'. As far as treatment was concerned, Hofer stressed the importance of keeping the mind occupied by something other than home – the need for companions 'by whom the imagination of the patient is distracted from that persistent idea'. In addition, he recommended cephalicum, mercury, opium, oil of hyoscyamus, purging pills, diaphoretic and stomachic mixtures, external cephalic balsams, and internal hypnotic emulsions to ease the sufferer's perpetual worries and spread a sense of warmth.

Above all, hope of returning home should be given. And if these measures had no effect, Hofer said that 'the patient should be taken away however weak and feeble, without delay, whether by a travelling carriage with four wheels, or by sedan chair, or by any other means. For certainly up to this time it has been proved by many examples that all those thus sent away had become convalescent either in the journey itself or immediately after the return to the native land; and on the contrary, many for whom means were lacking for a return to the native land, had gradually, with spirits exhausted, breathed out their life, and others had even fallen into delirium and finally mania itself.'

When Hofer's thesis was first published, it was widely believed that nostalgia only afflicted the Swiss. In German the condition was known as *Schweizerkrankheit* – the Swiss disease. In 1705, the Swiss physician J. J. Scheuchzer attributed nostalgia to the increase in atmospheric pressure experienced by these mountain-dwellers whenever they descended to the lowlands. Scheuchzer recommended that sufferers should be encouraged to return home immediately. If that were not possible, they should be sure to climb a nearby mountain or tower.

As more reports emerged, the idea that only the Swiss were

susceptible to nostalgia was gradually discredited. A footnote added to a 1779 edition of Hofer's thesis observed: 'The Scots, particularly the Highlanders, are also frequently assailed by homesickness, and the sound of the bagpipes which is very common among them can suddenly arouse this condition.' In a 1975 paper, 'Nostalgia: a "forgotten" psychological disorder', George Rosen refers to a treatise on military medicine published in 1754 by De Meyserey, physician-in-ordinary to the king of France, formerly a doctor in the royal armies in Italy and Germany. De Meyserey, observing how often nostalgia was brought on by tedium or vexations, emphasized the importance of keeping any soldier who showed signs of homesickness busy, diverted, occupied by tasks or vigorous activity. He recommended medications that would allow the blood and humours to circulate more easily. And, like Hofer, he insisted that, if these measures were not successful, the nostalgic patient should be allowed to go home, or at least be given hope of returning home.

It seems strange that nostalgia should have been so uniquely associated with the Swiss, when one of the founding works of European literature depicted so vividly the nostalgia of a Greek. When the reader of Homer's *Odyssey* first encounters Odysseus on Calypso's island, he is exhibiting symptoms which Hofer or De Meyserey would quickly recognize: sitting alone on a headland, gazing out over the sea, 'wrenching his heart with sobs and groans and anguish', pining for Ithaca – his loved ones, his high-roofed house, his native land. He spends his nights with Calypso, his days alone, sitting on the shore, 'weeping for his foiled journey home'. Later, he addresses those gathered in the hall of King Alcinous:

'How much I have suffered . . . Oh just let me see
my lands, my serving-men and the grand high-roofed house –
then I can die in peace.'

All burst into applause,
urging passage home for their newfound friend,
his pleading rang so true.

The *Odyssey* is only one of a whole collection of ancient Greek
stories (the *Nostoi*, or 'Returns') which describe the difficult
journeys of a character or group of characters back to their
homeland, especially the return of Greek heroes from Troy. These
stories endure because the pleading rings true: the reader under-
stands the strength of the longing for home, and appreciates the
deep affront of anything that foils or obstructs the journey back.
'Sunny Ithaca is my home,' Odysseus tells King Alcinous.

'Mine is a rugged land but good for raising sons –
and I myself, I know no sweeter sight on earth
than a man's own native country.'

*

AT SAND LAKE, snow geese had gathered on the ice in a great
assembly. Even Michael was bewildered: he'd never seen anything like it.

'Two hundred thousand geese,' he said. 'More even. Two
hundred and fifty thousand.'

We stood at the edge of the lake. Small groups of Canada
geese kept to the gold fringe of cattail and phrags. The ice was
covered with snow geese: a thick-sown crop of white necks, right
across the lake. Goose calls resounded in the ice, as if the hollow,
metallic din were trapped inside it. Sorties of geese took flight
from the assembly; squads returned from nearby fields, coasting
down on bowed wings and settling in the midst of the gaggle.
Suddenly, the flock took wing, an audience breaking into
applause. It was as if the ice itself had exploded – almost a
surprise to see the hard, blue-blotched plane intact beneath the

birds. The flock seethed, rolling back and forth on itself, its shadow roiling like a turbulence on the ice below. The applause deepened to the sound of trains thundering through tunnels. Scarves of glitter furled through the flock when drifts of birds turned their backs and white wings to the sun, and sometimes the entire sky was lit with shimmer, as if a silver, sequinned dress were rippling beneath a mirrorball, the sounds of goose calls and beating wings pounding the ice below. With binoculars I tried to follow individual birds through the pandemonium. I witnessed collisions – caroms and buffetings of blue-phase and white-phase snows, one bird's heft glancing off another's, the O of my binoculars a frenzy of black-tipped wings. Then, as before, the first birds settled on the ice, followed by others, each goose taking its place, the gaggle reforming bird by bird, the roar diminishing, until the whole flock, more geese than there are words in this book, was spread before us on Sand Lake.

'There's a group of them in the open water at the south end of the lake,' Michael said. 'Maybe forty thousand geese. If you park at Columbia Dam and creep through the phrags you might be able to get right up close to them.'

I drove to the dam and parked the Topaz behind a stand of elms. I walked through bromegrass towards the phrags. I could hear snow geese on the far side of the rushes, a low drone, ornamented by the grace notes of individual birds. The bright, straw-yellow phrags were seven or eight feet tall; each one had a feathery seedhead that brandished more than its fair share of sunlight. I stepped carefully through the rushes, trying not to make a sound, knowing that snow geese startle easily. The phrags thinned out, and I saw the geese, about twenty yards away. There were both blue-phase and white-phase snows, and a few of the smaller Ross's geese, which have the same plumage as white-phase snows, but shorter necks, and stubby, triangular bills. A few ducks, mallard and scaup, floated among the geese.

I crept forward through the phrags, as close as I dared. The seedheads glowed above me: it was a thicket of light. I didn't want to disturb the snow geese. Some of them had their necks up; they were just drifting, looking around. Others were sleeping, their necks turned and tucked down between folded wings, resting in the cradle of the back, as if the goose were both the nest and the bird inside it. Pairs and threesomes took off from the water; others came down to join the flock. The seedheads of the phrags quavered and sighed when wind blew across them. I stood still, watching the birds. Then I retraced my steps through the rushes and bromegrass to the stand of elms, hearing the chatter of the geese going on behind me. I got into the Topaz and drove back to Aberdeen.

*

SOME AFTERNOONS I'd drive back to Aberdeen and sit reading or writing in Dally's Dining Lounge on Logan Street. Dally's was glass-fronted: the word *Dally's* was engraved on the streetside glass in an ornate, slanting script. Customers took one step up from the street into a glass porch, then pushed open a glass door, surprisingly heavy, most people putting their shoulders to it. Inside were red leatherette booths, sprung like mattresses; a long counter with a row of stools; four slow-turning wooden ceiling fans; and, on the walls, black and white photographs of Aberdeen before the war: stanchioned streetlamps; men wearing cloth hats and coveralls; imposing, civic-minded clocks. Behind the counter was a cold cabinet of cheesecakes and deep-dish blueberry, pecan, cherry, lemon meringue and rhubarb pies, with tilted mirrors allowing patrons a bird's-eye view of the pies. There was a stainless steel Bunn-O-Matic coffee urn; a Silver King Imperial milk dispenser with a sticker on the front saying, *A Sign of Class . . . Milk in a Glass*; and platters of muffins and cinnamon rolls under scuffed Perspex domes.

The waitresses, Misti and Crystal, were in their twenties. They wore green-and-white-striped work frocks. Crystal had black hair furled in a bun at the crown; Misti's bleached-blonde curls tumbled off the escarpments of her shoulder-pads. They passed orders through a hatch to an invisible chef, and drank the dregs of milkshakes straight from the blender, priests knocking back the last of the communion wine.

Dally's closed every day at six o'clock. Misti and Crystal began to clean up at about half past five. Instead of wiping the surfaces with cloths, they pulled on pairs of wide, padded cloth mitts. The mitts lent the cleaning an intimate character. Misti cleaned the red pommels of the stools as if she were walking down a line of boys, ruffling their hair. Crystal wiped the tables with long, smooth strokes, as if she were grooming ponies. They caressed the Bunn-O-Matic with such tenderness it seemed the prelude to an embrace, and rubbed the Silver King as if a genie lived inside it. The mitts transformed the work into a ritual of solicitude that bordered on the sensual.

One afternoon, not long before mitt time, there was an almighty crash. Everyone in Dally's looked up to see a tall man walking straight through the glass door, mistaking it for open air, falling to the floor in a debris of chips and shards. He lay motionless in the glass debris. He had grey hair, held in place by pomade. A baseball cap, dislodged by the fall, lay upturned next to his head. Misti and Crystal ran over to him. The man moved his hand. The first thing he did was reach out for the baseball cap. Then he raised his head. There was blood on his forehead. He put the cap back on: it said, *Culprit – America's Favorite Fishing Lures*. Misti and Crystal protested when he began to heave himself to his feet. There was a chinking as bits of glass fell from his arms and shoulders.

'Whew!' he said. 'That was quite an entrance!'

Misti took his elbow and led him to a booth like a blind man.

'I think we should call an ambulance,' she said to Crystal.

'You don't need to do that,' the man said. 'I'm OK. I'm a little shaken, but nothing's broken. If I could just bother you for a tissue or something for this cut here.'

'Would you like some coffee, sir?' Crystal asked.

'Just a glass of water, thank you,' he said.

We all watched him closely, as though he were about to topple.

'My apologies for this rude interruption,' he said. 'That was a stupid thing. A stupid thing to do. I'm sorry. Dumb as a sack of hammers. Straight through a door. Thought it was open. That was a stupid thing to do.'

'Are you sure you're OK?' Misti asked. 'We could get you an ambulance.'

'I'll be fine,' he said. 'I'll just sit here a moment. My, those windows are clearer than air!'

'We shouldn't clean them so well, right?' said Crystal.

'Right!' the man said. He held a tissue to his forehead. He brought it down and saw the blood on it. He shook his head.

'Dumb as a sack of hammers,' he said.

Misti had fetched a broom and was sweeping up the glass around the door. Crystal had pulled the mitts back on; she was buffing the Bunn-O-Matic.

'I ought to go home,' the man said, standing up slowly. 'I'm no use anywhere else! What do I owe you for the window?'

'Don't even think about it,' Misti said.

'All right then. I'm going home.'

The tongue of his belt hung loose from the buckle, lolling like a dog's tongue. He lifted the *Culprit* cap and smoothed his hair back with two brisk sweeps. He walked out through the door, which now *was* open air, and took care on the step down to Logan Street. Misti and Crystal set to cleaning again. Lights sparkled in the metal and plastic surfaces, commending the work of the women's hands.

*

WINDS BLEW HARD from the south in the first week of April. Snow geese began to leave the prairie sloughs of South Dakota, sailing north on the tailwind. I saw them from the window of Michael's pickup as we drove west from Sand Lake on a dirt road across the Dakota Plain. The flocks were hurtling: earlier, driving north, Michael had seen geese flying above him, outracing the car.

'That wind's so strong,' he said. 'The geese hitch a ride on it. It's a free gift. They just hold out a wing and raft right into Canada.'

Globes of tumbleweed bowled across the track, northwards. On both sides, tracts of prairie were ploughed, cropped to stubble, or left as grassland. Wind blew the fine loessial topsoil from fall-tilled fields. The tilth rose in dark, north-drifting clouds like the smoke off stubble burns; dust-devils spun out of it like dervishes. Michael turned on the headlights when we entered a soil cloud.

'He tilled that and tilled that last fall and left no cover on it,' Michael said. 'And that's the most productive soil. That's bad farming in just about every way you can think of. Sometimes you can't believe the things people do. And that soil blowing off is probably filling the road ditch full of dirt also.'

We emerged from the cloud. Gusts came at the pickup broadside: we were heading on a beam reach, a prairie schooner; we keeled a little in the strong winds. The dirt track, rising slightly, scored an impeccable line ahead of us to the horizon. Abandoned grain elevators stood by the track, tin panels hanging from cross-braced timber frames. Prairie rolled out on all sides. Clouds raced northwards with the geese, as if to cloud breeding grounds. Sometimes, through gaps in the cloud cover, sunlight flooded the open country. Winds were starting to break up the ice on pothole lakes, driving wreckage in thick crystal flotsams against the northern banks, raising whitecaps on the water. We caught whiffs of a pungent, rotten-egg smell: hydrogen sulphide produced by vegetation rotting under the ice, suddenly released by the thaw. It was the land's exhalation, as if the prairies had held their breath all winter. Michael took deep breaths of it, filling his lungs.

'I love this smell,' he said. 'It's a sign of spring to me. That's why I like this smell so much. I *love* this smell.'

We crossed the Elm River. We passed the Leola Country Club, where a few incongruous dark green cedars betrayed the landscaping of a golf course.

'Leola's the Rhubarb Capital of the World,' Michael said. 'They have a three-day rhubarb festival here each June. You can get rhubarb fixed in more ways than you can imagine. There are a lot of abandoned farmsteads out here, and the only sign anyone was ever living there are rhubarb plants which just keep on growing.'

Between Leola and Eureka we passed a hill: a prairie knob. On the side of the hill, white rocks had been arranged in the letters LHS, which stood for Leola High School.

'Next week,' Michael said, 'it'll be EHS, for Eureka High School. Kids drive out from one town or another and shift the rocks. Back and forth. They just rearrange the stones. I remember once the stones had been laid out to say BEER ME, and once they said 69 ME, which is when we thought enough's enough, and rolled the rocks downhill. Came back a couple of days later and there they were again: LHS or EHS, the same old business.'

The prairie was less flat now, billowing like a sheet thrown out across a mattress: low, rolling grassland hills all the way to the Missouri River. Dirt plumed from the pickup's back wheels.

'This is the coteau,' Michael said. 'Beautiful country. Most of what you're seeing are grassland easements. That's land the government has purchased with Duck Stamp dollars and set aside from farming. Hunters have to buy Duck Stamps before they get to shoot waterfowl, and that money's used to protect breeding and staging areas for ducks and geese. Isn't this beautiful? This is how I imagine it used to be. No trees. Just grasslands and wetlands from the James River to the Missouri River. Prairie fires, bison and wind to stop the trees from getting a hold. If we were driving out here three hundred years ago, there'd be bison, elk, antelope, grizzlies and wolves, and no raccoons, no white-tailed deer, no red fox, no pheasants. We've changed it. Everything is so different. I'd love to have seen this country three hundred years ago. Used to be a lot of spotted skunk, but they're pretty much gone. One per cent of the original tallgrass prairie's left in the United States, but in South Dakota we've got six or seven per cent so we're doing pretty well. You can't recreate prairie. Once it's gone, it's gone. It's gone for ever once it's ploughed.'

The coteau had stone colours in this heavy grey light: ash-white, dun, tan, fawn. In the grasslands were traces of Native American camps, spots where bands of Dakota Sioux had made temporary homes amongst the Indian grass, panic grass, switch-grass, wild rye, sprangletop, dropseed and needle-and-thread. Their tepees, like coffee mugs, had left rings.

'The colours have bleached out over the winter,' Michael said, 'but in the fall I think the colours of the prairies, the golds, match any colour of woodland, even the fall shows of Vermont and the Appalachians. I've been over in New England when the maples are turning scarlet but I'll take the prairies any day. Sunrise when light's on the native grasses. The wind runs gold through the little bluestem. Isn't this beautiful? All you can see is prairie. No trees. I love country like this.'

I looked at Michael. He was looking from the road to the open prairie on both sides, his eyes fervent behind steel-framed glasses – a man in his element, inhabiting just the life that life intended. Close to the track, on our right, one hill rose above the hillocks, a distinct prairie knob, a prominence in the low, undulating land.

'We could stop for a moment,' Michael said. 'Might be nice to take a walk up there.'

He pulled over on to the shoulder of the track. He had to push hard to open his door into the wind. We put on coats and climbed barbed wire, one hand on the pommel of a fence-post. The wind ripped past our ears, gusting up to forty-five miles per hour. The grasses threshed and flailed. There was the roar of the wind in our ears, the flag-like snap and putter of our coats, the deep swishing of wind through stalks, leaves and seedheads across the coteau. We had to shout to make ourselves heard.

'Bromegrass!' Michael yelled, pointing at the grass around our feet. 'Kentucky bluegrass! Little bluestem!' I walked close to him, wanting to hear the names. He emphasised the names, pitching them against the wind. 'Western wheatgrass! Echinacea! Leafy spurge!'

We reached the slope of the prairie knob and made for the summit, wind driving at our backs. The wind hit the slope and accelerated, racing across the summit at sixty miles per hour, its force gathered on the long sweep of the Great Plains. All around us the prairie receded, falling away on the curve of the hemisphere. The wind was part vandal, smashing the ice in marshes and sloughs, and part thief, vanishing northwards with birds in its pockets. We leaned back into it, trusting our weight to the windspeed. Michael had spread his arms out wide, for balance, like a skydiver. He was lying back at close to forty-five degrees, and he was laughing, but I couldn't hear it: the wind filched his laughter along with the birds. Still reclining on the wind,

Michael pointed skywards. I looked up. Snow geese were flying high overhead, not arranged in orderly Vs and echelons, but tumbling in loose clusters and pairs, careening on the gale, not entirely in control of their own passage.

The following day I would pack my bags in the white motel room and drive the Topaz north, first to Fargo, then on across the Canadian border to Winnipeg. A few miles out of Aberdeen I would see flocks of snow geese flying *southwards*, working hard into the wind – geese that had hurried into North Dakota, encountered storms and inclement temperatures, and were now undertaking retreat migrations, fighting their way south to stay inside the scope of spring. But when I stood on top of the prairie knob, with the glorious grassland swales of the Missouri coteau all around, everything airborne was travelling northwards: winds, clouds, geese. The sky itself seemed intent on Canada. Michael and I lay back as far as we dared and watched the snow geese go, each of us laughing inaudibly to the other.

5: RIDING MOUNTAIN

IN 1763, JEROME GAUB, a Dutch professor of medicine and chemistry at the University of Leiden, related homesickness to unrequited love. 'How the force and continuity of the functions slacken,' he wrote; 'how the condition of the body languishes and all powers of the economy weaken and collapse when an ardent wish for some desired object is too long drawn out! Do not even the sturdiest races exhibit men who are troubled by peculiar ailments when assailed by a yearning, to which they do not yield soon enough, to return home after having tarried overlong in foreign parts, ailments that may end fatally when all hope of return is lost? How often do beautiful maidens and handsome youths, caught in the toils of love, grow ghastly pale and waste away, consumed by melancholy, green-sickness, or erotomania, when delays occur or the hope of possession is lost?'

In the late eighteenth and early nineteenth centuries, nostalgia was widely reported in the European armies, especially among soldiers who had been conscripted or impressed. Military physicians reported 'epidemics' of nostalgia. Most doctors believed that people from rural backgrounds were more susceptible to the condition than those who came from urban environments. Baron Dominique Jean Larrey, Inspector of Health of the French armies under Napoleon, discussed nostalgia at length in his surgical essays. Drawing on observations made during the catastrophic retreat from Moscow, Larrey divided the course of the disease into three stages. First, an exaggeration of the imaginative faculty: patients thought of their homes as enchanting and delightful, and expected to see relatives and friends advancing towards them. Second, the appearance of physical symptoms: fever, gastric disturbance, 'wandering pains'. Finally, depression, listlessness, weeping, and sometimes suicide. Like Johannes Hofer, Larrey emphasized the importance of distraction to prevent the onset of nostalgia: music, recreation, regular exercise – anything to keep the mind occupied, to keep the patient's thoughts away from home.

Weir Mitchell, a doctor in the American Civil War, reported that 'cases of nostalgia, homesickness, were serious additions to the peril of wounds and disease'. In 1943, David Flicker and Paul Weiss of the US Army Medical Corps published a paper entitled 'Nostalgia and its Military Implications' in the journal *War Medicine*. 'The greatest single factor in waging successful warfare is morale,' the authors began. 'A most important factor in attaining morale among fighting men is the preventing or overcoming of nostalgia.' Defining their terms, Flicker and Weiss quoted from *The New International Encyclopedia* of 1905: 'Nostalgia is a feeling of melancholy caused by grief on account of absence from one's home or country, of which the English equivalent is homesickness. Nostalgia represents a combination of psychic disturbances and must be regarded as a disease. It can lead to melancholia and even death. It is more apt to affect persons whose absence from home is forced rather than voluntary.'

Writing during the Second World War, Flicker and Weiss reported many of the same features of nostalgia that had so intrigued military physicians during the Napoleonic Wars. They considered nostalgia to be 'a contagious disorder which may spread with the speed of an epidemic through a company or camp'. They saw cases of homesickness increase as circumstances became more challenging, observing that 'whenever a division is moved a great number of patients pour into the neuropsychiatric wards'. They were in no doubt as to the force of the condition:

'Men in order to free themselves from nostalgia are capable of committing any sort of infraction or crime that is not sociologically or morally abhorrent. With this fact the military police are well acquainted; so when a soldier is absent without leave the police usually are certain that if they go to his home they can there apprehend him.' And, like Larrey, Flicker and Weiss point to the tendency of the nostalgic patient to idealize his home environment. 'As Freud has often indicated,' they wrote, 'distance lends enchantment, so that one forgets the many unpleasantnesses of his home or usual surroundings and can think only of the more desirable aspects.'

Certainly, the acute nostalgia I had experienced in hospital was precipitated by challenging circumstances, and had taken hold when my mind, without the diversion of companionship, activity or amusement, had the opportunity to ruminate and linger. Recent studies have shown that homesick boarding-school children are more likely to report homesickness at night or first thing in the morning: the day's events, by offering effective competition for the children's attentions, ward off the longing for home. Hofer and Larrey were quite right to emphasize the importance of distraction. And certainly, my desire in hospital was to go back to a place of unequivocal comfort, characterized not by argument, difficulty or sadness but by a set of reassuring images: rooks, the Sor Brook, animals huddled together in curtain folds.

*

THE DARK PINK CARPET in my Winnipeg hotel room was blotted with black stains, and there was a threadbare patch one step in from the door where footfalls were concentrated. Shabby, skin-pink curtains, smelling of cigarettes, were not enlivened by floral motifs in different pinks, as rooms are not enlivened by vases of dead flowers. The wallpaper's fuchsia and coral stripes had almost faded to a single tone; the dresser's pink paint was peeling on the drawers. Exposed water pipes shivered in flimsy, half-screwed fittings. A broad window gave on to Ellice Avenue, and that night, when the storm swept in, snow blew across it in flakes as big and fluffy as goose down feathers. Winnipeg was white in the morning.

Winter weather. No snow goose would fly north into conditions like these. No wonder I had seen geese fighting back into the south winds, retreating from the Canadian border. The birds were waiting for spring. As soon as conditions improved, snow geese migrating up the Mississippi, Missouri and Red River valleys would converge on the grain fields west of Winnipeg. The fields of the Portage Plains were the last major staging post on the journey from winter to breeding grounds. Geese would pause here to rest and replenish fat stores while the thaw worked northwards ahead of them. Hudson Bay and the first nesting areas lay another 1,000 miles to the north-east.

I'd read about the great quantities of birds involved in what John Dewey Soper called 'the spring bivouac in Manitoba'. In 1942 Soper wrote of 'amassments', 'mobilizations', 'swarming legions of the air': the war in Europe furnished such military figures. 'Is it any wonder,' Major Burt Gresham had asked in the *Winnipeg Free Press* in 1936, 'that estimates of the numbers vary greatly; that observers meet with politely lifted eyebrows of doubt when they talk of geese in terms of six cipher figures? How many birds would you say you saw in telling someone of a flight where the leading birds were dropping into the lake while the tail of the flock was still out of sight? Then how many would there be in a series of such flights, say at about half hour intervals during the late afternoon to long after darkness had fallen?'

Violence came with such numbers. Farmers launched sky-rockets at night to frighten off the migrant hordes. 'When one of these great flocks go up off a black, ploughed field,' wrote one observer, 'the effect is as if a volcanic eruption had blown the whole field into the air.' Once, near Elgin, Manitoba, snow geese were seen flying north-east during an electrical storm. The flock, 300 yards wide and three-quarters of a mile long, was flying at about 180 feet. Witnesses described a flash of lightning, a thunderclap, an entire portion of the flock falling to the ground, struck dead. Other birds, temporarily stunned, dropped with them but revived just in time to soar away.

I'd hoped to move north with the geese from the Dakota lakes to the Portage Plains. But I hadn't guessed the extent to which the weather would determine the pace or fluency of the journey north. I'd seen the migration plotted as a clean, unbroken arc from Texas to Foxe Land, imagining the flight to be as easy and consistent as the line, a graceful curve from one point to another. But the arc's grace was a fiction. The snow geese were improvising their way from winter to breeding grounds, stage by stage, according to the cue of the season. Now I felt foolish. I had no freedom. I was shackled to snow geese. If the snow geese were waiting, I would have to wait.

*

DAVID LIVED ON the edge of Riding Mountain National Park, 150 miles north-west of Winnipeg. A friend had given me his address, and we had corresponded before I left for Texas. David was an outdoorsman with a degree in biology, and he seemed to like my idea of following the snow geese from Eagle Lake to Foxe Land. I called him from the pink hotel room and explained my predicament.

'You could come out to Riding Mountain,' he said. 'There's a cabin. It's pretty simple. But you could stay there until those geese get their act together.'

'I don't mind simple. How do I get there?'

'There's a bus. The Grey Goose.'

I caught the Grey Goose on Easter Day, leaving Winnipeg for plains pieced out in mile-square sections, each section quilted with fresh snow. Here and there you could see farm buildings in horseshoe windbreaks, and cylindrical galvanized grain elevators standing in the wide open like Cape Canaveral launch towers, and when the faint outline of hills appeared to the north it was as if you were seeing land from the deck of a ship – as if the sea's flatness had run its course. The hills of Riding Mountain marked the extent of the prairies. From here it was boreal conifer forest all the way to the tundra of the far north. The sky was clear blue above the spruce trees. It was below freezing, but there was no snow.

The bus stop was an isolated gas pump, out of order. I jumped up and down to keep my blood moving. I kept an eye on the clear blue sky, just in case there were geese. I could smell the trees. I'd waited for about twenty minutes when a brown Suburban drew up at the gas pump.

'David?'

'You look cold. Jump in.'

He was past sixty, a retired teacher with grey, mussed, curly hair, and striking thick black eyebrows like two stripes of tar. He wore jeans, a pale blue Columbia fleece jersey, and white socks in Eddie Bauer galoshes. He kept a tin of Copenhagen chewing tobacco on the dashboard of the Suburban, and every few minutes he reached for the tin, unscrewed the lid with the fingers of the hand in which he held it, and tucked a fingertip of the fibrous paste inside his cheek. He spoke very slowly, with long pauses; he seemed to weigh each noun and verb like a small bag of sugar. Sometimes he used a little two-syllable cough to fill his pauses, and soon I began to hear the cough as a word, part of his vocabulary.

'So. How are the geese?' he asked.

'I've left them behind,' I said. 'I think they've dropped back with the bad weather.'

'Geese can be sensitive that way.'

'This is the first time I've had all the geese to the south of me.'

'They'll be up pretty soon. Everything's still frozen up here. Soon as that ice starts shifting, you'll see geese.'

We were driving through forests of white spruce, dark and evergreen, and trembling aspens with slender white trunks – tall, straight, close-packed trees rising on both sides of the road like the walls of a canyon, with dogwood and shrubby willow growing in the understorey, perfect browse for moose and elk. Sometimes the trees made way for sumpy frozen ponds fringed with dun rushes, and sometimes mounds of branch, brush and mud rose up from the middle of the ponds – beaver lodges in which underwater entrances, nest dens, feeding chambers and access tunnels were excavated.

'Never walk on the ice near a beaver lodge,' David said. 'Beavers swim around it, underneath the ice. It'll always be thinner than you think.'

'OK.'

'Do you know what a beaver young is called?'

'No.'

'A kit. A beaver kit. A bear cub, a seal pup, a beaver kit.'

We turned off the metalled road on to a dirt track, passing through a gate at the top of a steep hill. The track led to a wide clearing in the forest, with a woodframe house, half-clad in cream-painted clapboard, standing at one end. Beyond the house, just visible through spruce trees, was a long lake, fringed with rushes, still frozen, the same hard, blue-blotched white as Sand Lake. Close-packed conifers grew right down to the edge of the lake near the house, while on the far side there was open country: gently sloping fields of amber-yellow grasses.

'Lake Timon,' David said.

Two dogs had been dozing in the yard. They sprang to their feet as the Suburban approached – Harley, a black labrador, flush with pedigree, and Sitka, a mongrel, half coyote, charcoal and ash-grey, with the sharpened senses and dramatic startle responses of a wild dog. They jumped up at us as we got out of the Suburban, wagging their tails, emitting happy whines that broke now and again into melodious, full-throated croons. David greeted them, squatting down, holding their heads in his hands, letting them lick his mouth and chin.

Harley and Sitka came with us into the house. David's wife and father-in-law were sitting at a varnished pine table in the kitchen, drinking coffee. There was a pattern of brown fleurs-de-lys in the linoleum. A metal strip made a strandline between the linoleum and the deep-pile blue carpet of the living-room, where broad windows gave out on to the gleaming ice of Lake Timon. Karen was much younger than David; she was tall and rangy, with wavy auburn hair. She wore black leggings and a long, baggy red sweater with two daisies on the front; she was browsing a paperback cookery book entitled *Hotter Than Hell*. Her father, in contrast, was stocky and compact, in a lumber-jack's checked shirt, and bifocal spectacles in steel frames. He had stiff, tarnished-silver hair and a matching moustache that looked like a wire brush: it could have cleaned the verdigris from an old copper pan.

David introduced his father-in-law as the Viking.

'Why the Viking?' I asked.

'All my ancestors are from Iceland!' the Viking said, as if he'd clinched an argument. He spoke in abrupt, gruff bursts. 'Iceland was settled by Vikings,' he said. 'Came from Norway in 874. Icelanders came to Canada. It's in the blood! Viking blood!'

'Have a seat,' said David. 'Make yourself at home.' He crossed the strandline into the living-room. I looked out of the window at Lake Timon.

'So where are the geese?' Karen asked.

'They should be here soon,' I said. 'They got held up by bad weather.'

'Geese!' the Viking exclaimed, slapping the pine tabletop. 'Jesus Christ! What the holy smoke are they up to, those geese? That's a wild goose chase! Geese, geese, geese! So many damn geese! Holy Christ!' He slapped the table again. 'Those godforsaken geese!'

'Hold on a second,' David said, coming back into the kitchen, holding one palm out towards the Viking, as if to ward off whatever the Viking was about to say about geese. 'I'm going to take you to the cabin,' he told me. 'You can get settled in. You can just walk down to the house whenever you want.'

'OK.' I said. 'We can talk about geese later.'

'Holy smoke!'

We left Karen and the Viking in the kitchen and got back into the Suburban. Further up the hill, surrounded by thick forest, stood a tiny cabin: a prefabricated hut, just a few metres square, set on a swell of open ground, high above Lake Timon. Years before, when David had raised a small herd of German Gelbvieh cattle, he'd equipped the cabin with a desk and filing cabinet and used it as his office, so that buyers wouldn't tramp mud into the house. The cabin walls were still hung with photographs of prize Gelbvieh heifers, polled and horned. These radiant, butterscotch cows stood in front of placards commemorating the 1982 Regina Agribition or 1986 Alberta Farm Fair, and David stood beside the heifers in each of the photographs, wearing a Stetson and tan leather jacket, pride clearly visible in his face despite the shadow of the broad-brimmed Stetson.

He left me at the cabin and I took stock. This was where I

would wait for the geese. The desk and filing cabinet had gone. There was a narrow bed against the far wall, a sink, a rickety round metal table, a wicker armchair whose back fanned out wide behind the shoulders, and a cramped, cupboardlike bathroom. The gas in the old Perfection heater ignited with a *whoof* whenever the thermostat deemed it time. On a sideboard next to the sink was a row of books, bookended by glass jars filled to the brim with plastic, metal, china and tortoiseshell buttons. The books referred to the construction of the cabin, such titles as *Getting the Most out of your Lathe, Getting the Most out of your Drill Press, Getting the Most out of your Band Saw and Scroll Saw,* and *How to Do More with your Power Router.* I selected *The Wilderness Cabin* by Calvin Rutstrum and sat down on the wicker chair. The chair winced: the slightest movement provoked the thin, plaintive wince of wicker weave flexing.

There was a photograph of the author on the back cover. Wearing a checked shirt and a neat felt hat, he was sitting beside a lake fringed with spruce trees, a lake very like Lake Timon. He'd built his own cabin overlooking the valley of the Saint Croix River, an hour's drive from Minneapolis, and lived there for six months of each year. His cabin was built of square timbers and quarried stone, each piece cut by his own hands. He'd incorporated pine panelling, plank flooring, and a fireplace with a stone chimney. Calvin Rutstrum was an evangelist for cabins. He celebrated the cabin's 'simple, elemental form in our complex modern civilization'. He loved pioneer cabins roofed with split shakes, their walls banked with dirt, their split-log doors held together by battens on the hewn flat side; he loved the tensile strength and rigidity of fir plywood; he loved squared timbers, dovetailed corners, flooring properly dressed and matched. He recommended fir plywood gussets as bonding plates, and liked to shim rather than plane down the uneven thickness of boards in order to preserve their original ruggedness. He usually built a small auxiliary cabin before starting work on the main dwelling. Once this was done, and after the first tattoo of rain sounded on the roof, he would get the feeling of 'belonging' to the particular country in which he found himself.

Sometimes I leaned back in the chair just to hear the wicker wince. It was getting dark. There was only one light: a lamp with a red shade, and pricks and incisions in the shade that seemed randomly distributed until you switched the lamp on and light shone through them in the likeness of a cat, sitting upright, with a ribbon round its neck, tied in a bow. There was another cat in the cabin, sitting on the window-sill above the round table – a kind of jack-in-the-box, a grey plastic cat leaping from a drum sashed in chintz, a rope zigzagging from one rim to the other. This cat wore a chintz dress, a collar of lace, a pink ribbon, and a conical hat of lace and chintz, topped with a white pompom. It had the oversized blue eyes of a Barbie Doll, and a pert, pink nose.

I went back to *The Wilderness Cabin*. 'Mental and manual exercise together shape happy lives,' Rutstrum argued. 'Much of our thinking, we might say, is actually done with our hands.' A few well-stocked shelves of food and equipment, a stove, a simple board table, some empty root-beer cases for stools, a cot or bunk bed, a supply of books, a portable galvanized bathtub, an all-band transistor radio – this was all you needed to feel at home, to belong to a place as it belonged to you. Reading *The Wilderness Cabin*, you understood there'd be difficulties (cabins settling with log shrinkage or keeling on the frost heave), but these seemed insignificant next to the music of tools and techniques: the bull points and scutches, the tin shears called snips, the augers and gimlets for boring holes in logs, the splines for window-frames, the lug hooks and peaveys for holding logs firm when sawing, the India oilstone for honing an axe's blade, the cleaving tools called froes for splitting rough shingles or shakes from billets of timber.

It was a song of home-making. There was no shortage of logs, Rutstrum insisted: logs grow on trees. 'Outside doors to bedrooms enable early risers to depart without disturbing the household,' he observed. 'Where separate cabins are the choice, small kitchenette areas enable your guests and you to enjoy those periods of privacy so important in prolonged human relationships.' It was dark now. I read by the light of the lamp, a magic lantern, the cat cut into the red shade. 'What is ever more important to us,' I read, 'is our closer awareness and our richer understanding of nature, of the flight of birds, the changes in the sky, the pattern of the stars. These wonders, contemplated from day to day, will bring us the peace from which we have been too long estranged.'

The wicker chair winced. Wolves and coyotes were howling. The cabin was surrounded by Riding Mountain's wilderness forest: black bears were hibernating in nooks and culverts; moose and elk were browsing willow shoots; beavers were weakening the ice around their lodges, preparing dens for the summer's kits. David's woodframe house was half a mile down the hill. I was grateful to him for this shelter. I had stolen a march on the snow geese and would sit tight for as long as it took. I returned *The Wilderness Cabin* to its place between the jars of buttons and turned off the lamp, extinguishing the cat.

*

IN MARCH 1875, the Dyngja Mountains of Iceland exploded in volcanic eruptions. Pumice covered an area of 2,500 square miles, two or three inches deep, devastating farms. Many Icelanders had no choice but to emigrate, and some chose Canada, following the lead of Sigtryggur Jonasson, who had left for Quebec in the summer of 1872. The settlers stopped first at Kinmount, Ontario, then pushed west, drawn by what they had heard of the Red River valley. Its soil (black mould over white clay) was rumoured to be extremely fertile, and the region was said to support an abundance of ducks, geese, moose, strawberries, raspberries, blueberries and currants. Lake Winnipeg, an inland sea more than 300 miles long, promised a boundless supply of whitefish, sunfish, catfish, pickerel, sturgeon and pike.

The journey from Iceland to New Iceland took six weeks. Almost 300 settlers travelled by steamer from Sarnia to Duluth with all their worldly goods, hogs included. They boarded a train at Duluth and rode in boxcars to Fisher's Landing on the Red River, where they took an old stern-wheeler steamboat, the *International*, north to Winnipeg, transferring to flatboats to navigate the shallows, and transferring again at Stone Fort to a dilapidated Hudson's Bay Company steamer, the *Colville*, for the last leg to Willow Point, which was right up near Lake Winnipeg.

The first town built in New Iceland was named Gimli, which was the name of the home of the gods in Norse mythology. Gimli means 'great dwelling' or 'great abode'.

'They had trouble catching fish,' the Viking told me. 'They were deep-sea fishermen, most of them, and lake fishing's a whole different can of beans.'

We were sitting at the glossy pine table in the kitchen, feet on the fleurs-de-lys.

The Viking's grandparents had helped build Gimli. 'My father was a stonemason,' he said. 'Loved music. Could pick up anything and find a tune in it. Played tuba in the Argyle Brass Band. Best band in the country, some say. Played concerts, sports meets, exhibitions. Played at Baldur, Glenboro, Cypress River, Holland, Greenway, Belmont. And I'm tone deaf. My father could knit, too. When I was young, in the evenings, my mother and father would knit. My mother could knit and play cards at the same time. My father was a good knitter. He could knit a pair of socks in a couple of days, but he had to give it his full attention.'

Usually, when I walked down the hill to the house, David would be in his office, working on the computer, Karen would be teaching in Erickson, and the Viking would be fixing things. He'd have his head inside the drum of the tumble dryer because the belt was running awry on the idler pulley, or the hood up on his truck so that he could check the coolant level in the radiator or the condition of the drive belts. Or he'd be in the bathroom with the top off the flush tank, adjusting the float ball. He'd be brandishing hammers, screwdrivers, pliers, spanners or wrenches, with screws or nails held between his teeth. He'd be wearing a checked shirt, and jeans held up by braces and a black leather belt. Canadian folk tunes would be belting from a paint-spattered radio-cassette. The Viking loved to listen to a fiddle player named Mel Bedard.

'I've heard Mel Bedard play in a dancehall for twelve hundred people,' he said, raising his wrench for emphasis, 'and the more he sees them dance, the more he bends that bow. He *makes* them dance: people who've never danced before, once they hear Mel Bedard they just can't help themselves!' He reached with his left hand for the neck of an imaginary violin, and played along with the tape using the wrench as a bow.

One morning I walked down to the house and found the Viking holding court in the kitchen. Two other men of Icelandic descent, both in their sixties, were visiting. They were called Bjornson and Bjornsson.

'We're not related,' Bjornson said. 'He's got one more "s" than I have.'

The three men were drinking coffee at the pine table. Bjornson had taken off a blue baseball cap and placed it on the table beside his mug, peak folded inside the body of the cap. He had thin, light brown hair and a wide, flat face. He was soft-spoken. He'd been a farmer: a small herd, some wheat. Bjornsson was heavy and garrulous, with a fulsome three-ply chin and strong Icelandic accent. He wore the kind of sepia-brown Polaroid glasses favoured by anglers, and a stone-coloured multi-pocketed vest such as Rollin had worn at Sand Lake. Underneath the vest was a black T-shirt printed with the legend *Be Staunch. Walk Tall.*

'I made butter for forty years,' Bjornsson said, rocking back and forth on his chair, which seemed very small beneath him, his eyes indistinct behind the Polaroid lenses. 'Government shut me down. Regulations! We bought the best cream from the dairy farms. Now they put all other kinds of things in butter, all kinds of milk products. This whipped butter, do you know what that is? Air! It's just air! They whip a lot of air in it, so it'll be softer, but you get less butter. It's like ice-cream. You get good hard ice-cream and the soft-serve ice-cream all the young people are eating, and you know what makes it soft, the soft-serve? Air!'

The Viking wasn't listening to Bjornsson.

'Iceland!' he said. 'There's a place!'

He'd gone to Iceland for the first time the previous year, on a tour, wanting to trace his relatives. The bellboy at the hotel in Reykjavik had turned out to be his second cousin, and when the Viking had asked the receptionist to look someone up in the telephone directory, the girl had said that wouldn't be necessary, they were best friends, she knew the number by heart. What had struck him most was the greenness of the country.

'Everyone thinks it's all snow and ice and glaciers, but my god that place is so green!' he said. 'All that grass! Meadows! And cows grazing, everywhere!'

'The cows wear brassières,' said Bjornson.

'That's right!' said Bjornsson. 'They've got cows in these big leather brassières. Stops the udders and, what are they, *teats* from knocking on the rocks!'

The three men rocked back on their chairs, guffawing.

They began to talk about the Gimli parade. Each year, on Islendingadagurinn, or Icelanders' Day, Gimli hosts a festival whose centrepiece is a parade of floats. A woman stands on the leading float, representing the Fjallkona, Maid of the Mountains, symbol of Iceland, dressed in a white gown, a green robe trimmed with ermine, a gold belt, and a high-crowned headdress of white veil falling down over her shoulders to the waist.

'One hell of a beautiful lady,' the Viking said.

'She is indeed,' said Bjornsson, sombrely.

'That's the truth,' added Bjornson.

Once or twice, I learned, the Viking had ridden on the leading float next to the Fjallkona. Bjornson and Bjornsson encouraged him to bring out his costume. The Viking disappeared for a minute and came back wearing a bizarre helmet: a yellow microwave-compatible cooking bowl, with two curving cardboard horns attached to it with tape, and a red leather key fob taped to the rim of the bowl, hanging down over his nose.

'See these?' he asked, pointing to the two horns. 'I'm the horniest Viking in Manitoba!'

The kitchen swelled with laughter. Bjornson and Bjornsson rocked back on their chairs. Bjornsson rocked back an inch past the balance point and only just managed to grab the corner of the sideboard to stop himself from falling. The room's laughter took a breath, saw Bjornsson was safe, then surged again, with renewed force.

I noticed the Viking's belt buckle. He was standing in front of us in the horned helmet. His buckle was a silver plaque embossed with the front ends of a trolley bus and streetcar. For ten years, after the war, he'd driven buses and streetcars in Toronto.

'Drove the Long Branch routes,' he said. 'And Neville Park, Dundas, Humber, Spadina Station. Toronto was all busy, busy, busy. Crowds, crowds, crowds. Rush hour lasted three hours. Moved to Winnipeg in 1955. The streetcar service ended in Winnipeg in that year, 1955. Drove a brand-new fleet of trolley coaches from Canadian Car & Foundry. Drove Sargent, Notre Dame, Logan, Ellice, Salter, Polo Park. Then they replaced the trolley coaches with diesel buses. Killed the DC machine at Mill Street substation. First time that place had been quiet in I don't know how many years.'

Behind him, against the wall, was an old wood bookcase with four painted duck decoys on its shelves. Remembering Michael, I identified the species: mallard, bufflehead, canvasback, pintail.

'Bought it at an antique sale,' the Viking said. 'Got it for peanuts, on account of the marks all down the side here.' He pointed to an end panel. The wood was notched with short horizontal lines in blue ballpoint pen. Not one of the lines ran quite true, the ballpoint working across the grain. Each line was accompanied by a date. I recognized the marks: a child had stood against the bookcase to be measured. I thought, instantly, of home, the white wall in the bathroom, beside my mother's grey heron, our heights inching their way towards the ceiling, heels to the skirting-board.

That afternoon the storm reached Riding Mountain. After a couple of hours snow lay thick on Lake Timon and the forest tracks. In the evening, when I set out from the house up the hill towards the cabin, snow was still driving. There was no moonlight. The track was a faint white band between impenetrable black woods. I trudged through the snow, head tucked low, my body angled forward against the wind like a letter in italics.

Snow had drifted against the cabin door. Once inside, I switched on the lamp. The cat appeared, the red shade's trick of pinholes and slits. The *whoof* of the Protection's gas igniting startled me. The cabin creaked and shook in the wind. It seemed the storm might wrest it loose and set it down in Oz.

The grey plastic cat was still bursting from the drum on the window-sill. I tried to avoid its stare. At the back of the drum there was an old-fashioned clockwork key. I turned it; I heard it ratchet up a spring, click by click. When I let go, a mechanism inside the drum was set in motion, producing a quaint, pinging, musical-box melody, causing the cat's body to sway and twist, its arms moving up and down, reconfiguring the shadows in the chintz gown. The melody slowed as the clockwork ran down; the pauses between pings got longer; the cat's dancing became a sequence of dreamy throes, as though it were dancing in deep water, its eyes remaining wide, unblinking.

The music stopped; the cat froze. Wind blustered through the trees, rocking the cabin in sudden surges. I imagined myself inside a tiny box of light in the middle of the forest. I thought about snow geese, still held up in the south by winter weather. I wondered how long I would have to wait until we were under way again. A gust caught the cabin, shaking it, tripping the drum's clockwork. The mechanism played three last unexpected notes. I turned quickly, shuddering, the cat's head moving through one last quarter-turn, snow driving across the window behind it, rose-lit by the lamp, the howls of wolves and coyotes no longer audible for the wind. I undressed and got into the narrow bed, shivering, curling up, thinking of home.

*

THE SWIFTS CAME BACK each year, in the last week of May. These were common swifts, *Apus apus*, sooty black all over save for a pale chin, known variously as skeer devils, swing devils, jack squealers, screech martins, shriek owls or screeks – names that alluded to the bird's fiendish screaming flight and diabolic black appearance. Swifts like to nest in nooks in the stonework of high walls, under eaves, even among rafters, and show a high degree of philopatry (from the Greek words *philein*, 'to love', and *patris*, 'homeland'), with generation after generation returning to favoured nesting sites. The advantages of this behaviour are clear: if a bird is familiar with its environment, it is likely to be less susceptible to predators and more efficient at finding food. Philopatry tends to develop in species that nest in stable, reliable sites such as cliffs or buildings, rather than in species that use unstable sites like river sandbars. There's no point returning to a place if you can't rely upon its qualities.

Every year swifts returned from Africa to the medieval ironstone house in the middle of England, making a beeline for the eaves on the warmer south side, the stone slate roof tiles colonized by lichens in tie-dyed coronas of white, rust and pale green. My father would expect them like dinner guests on 23 May, and they would stay at the house through June and July, until at the beginning of August they set off for Africa, fledglings leaving for winter grounds several days before their parents. A juvenile swift, like a juvenile cuckoo, depends on its inherited, endogenous migratory programme and compasses to lead it south through Europe to western Africa. From there, common swifts filter gradually southwards with the inter-tropical weather front (a confluence of airstreams that draws up insects from sub-Saharan Africa and dissipates at the Gulf of Guinea in late autumn) before turning east across the continent. In the winter, swifts ringed in Britain are most often found in Malawi. One in six perishes on the way.

Common swifts have long, thin, recurved wings, and short bills with wide gapes, evolved for catching airborne insects: they feed on the aerial plankton of aphids, beetles, spiders, hoverflies, leaf-hoppers, crane flies, spittle-bugs and thrips. David Lack, who studied swifts nesting in ventilation shafts in the tower of the University Museum in Oxford, found that a swift brings just over a gram of insects to its young in each meal. One pair brought forty-two meals to their brood in a single day – a load, Lack estimated, of around 20,000 insects. The warm, sheltered air on the south side of the ironstone house was full of insects, and in June and July, at dusk, parties of swifts exploited this abundance, breeding birds darting again and again to the eaves, delivering bugs to their nestlings.

In the cabin on Riding Mountain, with the storm blowing outside, I thought about those displays. The little hut kept creaking and shaking. I was wearing thermals and socks, with my coat spread on the narrow bed as an extra blanket. Eyes closed, I remembered how I'd sit out at the back of the house after supper, watching swifts, not long after I'd found *The Snow Goose* in the hotel library and begun to pay some attention to birds. Two or three months had passed since my last spell in hospital, and my attitude to the house had started to change. I felt angry at my prolonged confinement, desperate to be back in the world, as if my childhood home were somehow separate from the world, a zone apart. I escaped myself when I watched the swifts. And now, with the storm swirling round the tiny cabin on Riding Mountain, the memory of those evenings was itself a kind of retreat or sanctuary from surrounding turbulence. Rooks were calling raucously. A light wind swished in the trees like crinoline dresses on a ballroom floor, the *trespasses* sibilance of the Sor Brook going on beneath the crinoline dresses. I heard the back door opening (the bars of three bolts slid with known weight and easiness into sockets on the jamb) and saw my father walk out on to the lawn, shirt-sleeves rolled up, holding a mug. He stood beside me, and we both looked up.

Furious activity in the twilight. Eight, ten, twelve swifts were wheeling overhead, black birds racing round and round, their trajectories ornamented with swoops, tilts, rolls, dips, glides and zigzags, and rapid shimmies as individuals diverged from a straight course to take an insect seen to one side. The swifts feinted in one direction only to curve off in another, five or six birds appearing suddenly in formation, a squad, their stiff wings held fast in sickle arcs to carve a turn, then beating again to crest the roof-slope, tails opening in two-pronged forks for increased control when manoeuvring, then closing, streamlining the body for fast flight. The swifts flew like blades, birds slicing in and out of the paths of other birds, their shrill, sweet screams intensifying and fading in quick Doppler shifts as they passed overhead – accelerating, tipping, flirting with angles, leaning into banked turns that seemed to scour out the inside of a sphere. Sometimes one swift flew alongside another, their speeds, curves, shimmies and feints matched with unfailing exactitude, as if every movement were plotted by a common whim. Rooks flew west over the house in feeding sorties, heavy black rags adrift in the dim light, their flight sluggish and laboured next to that of the fleet, trim, screaming swifts.

My father delighted in these aeronautics. 'There's joy in it,' he said. To him, the return of the swifts was cordial and fortifying, a sign that the centre was holding, that orbits were regular and true. The arrival of migrants, like an eclipse, was a revelation of planetary motion. The Earth had travelled once more round the sun. Circuits were correctly aligned. Seasons were respecting their sequence. Time could be relied upon.

The day was going. We watched the swifts on their precipitate vespers flights. The swift family has evolved almost exclusively for an aerial existence, and the needle-tailed swifts of Africa and Asia are the fastest of all birds in level flight, capable of attaining speeds of up to 105 miles per hour. Swifts have strong claws for clinging but aren't well-fitted for perching or walking. They have small, inconspicuous legs; their name, *Apus*, comes from the Greek for 'without feet'. A swift spends almost its entire life on the wing. Unless forced down by accident or storm, swifts stop flying only when they nest. They take nest materials from the air – the sky flotsam of leaf and chaff – and drink by descending in a shallow glide to open water, sipping as their heads touch the surface and shivering as they rise to shake any water from their feathers.

Swifts bathe in the rain: they take showers. They mate on the wing and even sleep on the wing, the feathers on their lower leg bones keeping them warm at night. A French airman in the First World War, recalling a night reconnaissance mission on the Vosges front, described how he had climbed to 14,500 feet above French lines, then cut the engine and glided down over enemy territory. 'As we came to about 10,000 feet,' he wrote, 'gliding in close spirals with a light wind against us, and with a full moon, we suddenly found ourselves among a strange flight of birds which seemed to be motionless, or at least showed no noticeable reaction. They were widely scattered and only a few yards below the aircraft, showing up against a white sea of cloud underneath. None was visible above us. We were soon in the middle of the flock. In two instances birds were caught and on the following day I found one of them in the machine. It was an adult male swift.'

At about nine o'clock, the screaming parties at the back of the house would start to gain height, rising gradually, disappearing just before dark, heading for the thinner air of high altitudes, where less energy was needed for flight.

It was that time now. We couldn't stop watching the swifts. My father was mesmerized by their flights, their courses twinkling with feints and shimmies. He held the mug close to his chest, without raising it to his lips, swifts whirling round and round above him, impelled by stiff, curved wings, fiendishly deft and spirited, tailed by screams like fine silver streamers. They began their night ascents, wheeling high above the roofline. The rooks stopped cawing. I got to my feet and leaned back against the house, surprised – thrilled – that the walls should be so warm, steeped in the hot day, remembering it. I stood with the warmth of the stone in my shoulders, pressing my palms against the house. I looked up, but the swifts had gone. 'That's it,' my father said. He turned, pushed open the back door, and stepped inside.

*

THE STORM PASSED OVER Riding Mountain; the thaw came hard on its heels. Snow melted quickly, draining into the sloughs and marshes. Snow beds and banks lingered in shaded places. Every day I took the two dogs walking, hoping to see snow geese. Harley, well-bred, obedient, stayed close to me on the track, while Sitka, ash-whites and charcoals blended in her fur, rummaged wildly in the undergrowth or bounded far ahead, dashing up to vantage points and striking a pose, all instinct, sizing up the territory. About a mile from the cabin the track emerged from the forest to a rousing prospect of open, rolling country covered with wheat stubble and blond prairie grasses. Ice was breaking up on the sloughs: Canada geese, white chinstraps agleam on jet black necks, joined mallard and goldeneye on the open water. A red-tailed hawk glided low over the grass. A bald eagle perched on the point of a spruce. A great blue heron flew down in front of me to take up position at the edge of a pond, its wings making the *whup-whup* of someone walking in a sarong.

Birds were coming in. I looked for snow geese every time I went outside. Migrants appeared on the birdtable outside David's kitchen window, a new species every day: red-winged blackbirds, dark-eyed juncos, evening grosbeaks, purple finches. The Viking looked forward to later arrivals, especially the Baltimore orioles and ruby-throated hummingbirds. He raged at burdock, a weed growing in the sere grass around the house.

'Burdock!' he ranted. 'It's the curse of the world! It's the bane of my existence! Holy Christ! This burdock!'

Fluff from cattail seedheads skiffed past him on a breeze.

There was still no sign of snow geese. Each morning I sat writing at the cabin's rickety table, under the eye of the chintz-gowned cat, and in the afternoons I walked with the two dogs along tracks through the spruce forest or on long circuits around Lake Timon, often singing or whistling, wary of the black bears that would be emerging from hibernation as the days grew warmer. 'Make a noise,' David had advised. 'Bears don't like to be surprised.'

One afternoon I walked up the track from the cabin with Sitka and Harley. We came out of the trees, and the dogs ran ahead, bounding through the wheatgrass. Cloud shadows roamed across the open country. We struck out through knee-high wheatgrass towards an old homesteaders' cabin, tilted like a trapezoid, its logs chinked with mud, half the shingles missing from the roof, the floorboards rotten. Tatters of wallpaper fell to the floor when I went inside, the dogs already chasing rats in the root cellar. An old leather shoe rested on one of the few remaining floorboards, hardened and gnarled, like a twist of driftwood. Bedsprings were strewn about. The stove was intact: a Peninsula Monarch, made by Clare Bros. & Co. Ltd of Preston, Winnipeg and Vancouver, with a dial showing the baking heat, *Warm* to *Very Hot*, and a rusty spoon lying across the hob. The cabin was disintegrating, but the stove – the hearth – was resolute, unshiftable, apparently resistant to the processes of decay.

On the way back to the house, the dogs caught a muskrat and tore it clean in half. They trotted along, holding their halves of the muskrat in their teeth. Harley had the head and chest, the rodent's arms protruding from the corners of her mouth; Sitka had the haunches and rear, the muskrat's tail swinging loosely below her chin. We found the Viking cleaning his pickup, swabbing the hood and windscreen with soapy water and a brick-shaped yellow sponge, dressed in a denim shirt buttoned right up to the collar, and jeans of the same light blue. His jeans were held up by braces in addition to the black leather belt with the silver buckle. One, it seemed, was a back-up system for the other.

'You!' he shouted as we approached. 'Have you looked up lately?'

'Where?' I said.

'Right here! Holy Christ! Look!'

He pointed. I looked straight up at the sky. Snow geese were flying overhead, blue-phase and white-phase birds, three distinct Vs, coming from the south.

'What took you so long?' the Viking yelled at the geese, shaking his sponge at them. 'You're late! Jesus Christ! You got people waiting for you! Holy Christ!'

The dogs put down their muskrat pieces and looked up.

*

IT WAS THE MIDDLE of April. I'd planned to see the snow geese on the Portage Plains, then take the train from Winnipeg to Churchill on Hudson Bay. The Viking, who was due to go back to his apartment in Winnipeg, offered me a lift.

'We'll find geese on the way,' he said.

Harley and Sitka ran after the pickup, barking. We drove out of the spruce forest, back down off the hills of Riding Mountain to the open plain. The Viking had fitted a set of flip-up shades to his steel-framed glasses. He flipped the lenses up and down indecisively. He drove slowly but pressed himself right back in his seat, arms at full stretch, as though experiencing substantial g-force. I kept an eye on the fields and sky, looking for geese.

Before the war, the Viking had worked for a year as a baker. He joined the army at nineteen and landed at Normandy the day after D-Day, a frontline signaller, carrying a radio on his back. After the war, he'd wanted to be a jeweller.

'Why a jeweller?'

'Because I've always loved tinkering with things and handling little tools.'

'What happened?'

'At that time the government was offering lots of training opportunities. So I said I wanted to be a jeweller. Government told me the jeweller's courses were reserved for disabled veterans. End of my career as a jeweller. Joined the transit company in Toronto. Drove trolley buses and streetcars. Met my wife. Got married. She was a real box of tricks. We'd both grown up in Manitoba, so we moved back to Winnipeg. I'm divorced. Had my camera ready when she left the house with all her suitcases. Wanted a picture of her going out through that back door for the last time. She saw me standing there with the camera ready. Picked up a mop and hit me in the face. Holy Christ!'

We were driving south across the flat land, the Viking pushed back into his seat like an astronaut. The plains were organized in mile-square sections of wheat, barley, corn, flax and canola stubble. Farm tracks and telegraph wires, crossing at right angles, ran along the edges of the sections. There were shelterbelts of green ash, and farm buildings just visible in windbreaks. We passed a cemetery.

'All the people in there are dead,' the Viking said. 'Every one of them.'

Sometimes we left the highway and pulled up beside lakes. All the ice had gone. I passed my binoculars to the Viking; he flipped up his shades and pressed his bifocal lenses to the eyepieces. We saw Canada geese, mallard and lesser scaup, and eight American white pelicans attended by double-crested cormorants and Franklin's gulls – small freshwater gulls with black hoods. They may all have flown up from the Gulf of Mexico.

Snow geese flew high overhead in undulating skeins and echelons. Flocks of killdeer – plovers with white underparts and two distinctive black breastbands – took off from stubble fields. I saw a V of long, lanky birds with heavy, slow-beating wings, necks stretched forward without a kink, legs trailing loosely behind them in bunches, and wondered if these sandhill cranes had flown from Texas – had even wintered on the prairies around Eagle Lake, stepping with the dainty gait of ballerinas along the edge of Jack's holding pond. I spotted a flock of large white birds in a field of canola stubble and asked the Viking to pull over, thinking they might be snow geese. But the white birds were tundra swans, much larger than snow geese, with black legs and bills, and all-white plumage, without the geese's black-tipped wings. Tundra swans winter in small pockets on the coasts of the United States and migrate to breeding areas from Alaska east to Baffin Island, across the far north of Canada. These swans in Manitoba were staging like the snow geese, gleaning for grain, replenishing fat stores, waiting for the thaw.

We drove on, in no particular direction, looking for flocks of geese. Monumental structures appeared on the level horizon to the west: sand-coloured, like half-completed pyramids.

'I know what they are!' the Viking declared.

Bales of flax straw, stacked in massive ricks. The bales would be sent to Pennsylvania for processing into cigarette papers and high-grade writing paper.

'Saw one of those damn things burning,' the Viking said. 'One summer. Lightning got it. Holy smoke, it blazed like a bonfire! Miles away, you'd see it, on the horizon. Column of black smoke like you wouldn't believe it if I told you. Jesus Christ!'

'When did you get to know all the birds?' I asked.

'Used to hunt all the time as a boy. I had a shotgun. One day I went out on my own, brought down fifty-two mallard, the whole pack of cards.'

'How did you get them home?'

'Oh, there were straw bales lying around. Took some of the twine off the bales and tied all the ducks together. Tied the other end to the seat of my bicycle. Rode home dragging fifty-two mallard ducks behind me, and, Christ, they were heavier than carpets.'

After his divorce, five years passed before the Viking spoke to his wife again. He saw her once during that period. He was on a shift, driving down Portage Avenue. He was waiting at a stop-light. He saw his wife in a car on the far side of the intersection.

'Do you know how that affected me?' he asked.

'How?'

'I actually had difficulty in breathing.'

We had been driving across the plains for almost three hours when we found snow geese on the ground. The Viking pulled over; we got out of the pickup. Perhaps 10,000 birds were gleaning in wheat stubble. Some were alert, their heads raised, periscoping; others were nosing in the black soil for leftover grain. A few tundra swans walked among the snow geese like samurai: grander birds, with more shining, imposing figures, a purer white. The calls of the geese combined in an insistent drone, graced with individual yaps, topped by the descant keening of killdeer. A flock of the plovers took off and settled again on the stubble between the snow geese and the track where I stood with the Viking.

'Hello, birds,' he said.

The Viking began to comb his hair. He kept two plastic combs in the back pocket of his jeans: a black comb, and a smaller pink comb with the fine tines of a lice comb. The way he drew, teased and flicked the combs through his tarnished-silver hair reminded me of his ambition to be a jeweller, his love for 'little tools'. He used the black comb for outline and general form, then switched to the pink comb for precision work and ornament. His hands were heavy, rough and seasoned, but he handled the combs with the quick-fingered dexterity of an illusionist.

Small parties of snow geese took off from the flock and flew away from the field; other parties flew in and took their places in the gaggle. The geese coasted down on bowed wings, dropping their legs like an undercarriage, their bodies tilting backwards, wings beating in reverse thrust, until at the last moment each bird's weight seemed to drag forward through its body, the feet touched down, the goose folded its wings and set to feeding. Herring gulls hung on the wind, crying like oboes.

The Viking returned the combs to their pocket. He was wearing his jeans and denim shirt. The shirt was buttoned up to the collar; the jeans were held up by braces and the black belt. The Viking was troubled by the attention of widows.

'In my building alone there are eight widows,' he said. 'You have to fight them off, I tell you. They're beating my door down. Holy Christ, these widows don't take no for an answer. These widows *terrorize* you. A man's not safe in his own home. Let your guard down for an instant and those widows *leap* upon you! Before you know what's happened, you're saddled with a widow. Holy Christ!'

The afternoon was beautiful: unambiguously spring.

'Look at that,' the Viking said. 'Wall-to-wall sky. My brother used to love those white fleecy clouds just sailing by. Look at that!'

The flock was never still. The geese shuffled across the stubble, blue-phase and white-phase birds intermingled, gabbling constantly, feeding or watching for predators. This was the key advantage of living in a flock: the more birds were gathered together, the less time any one bird would have to spend looking out for danger, and the less time any bird had to spend looking out for danger, the more time it would have for feeding. I wondered if I'd seen any of these geese before: on the prairies outside Eagle Lake at sunset, high over Matthew's half-built house in the hills, flying parallel to the Greyhound south of Minneapolis, or rising off the ice at Sand Lake, somewhere in those swirling crowds.

'So what are you going to do?' the Viking asked.

'I'm going to pick up some warm clothes in Winnipeg,' I said. 'Then take the train up to Churchill.'

'Then what?'

'That depends. When the geese start moving on from Churchill, I'm hoping to catch a plane up to Baffin Island. The biggest nesting grounds are said to be on Baffin Island.'

'Then what?'

'Then I'll go home.'

We turned away from the flock of geese and went back to the pickup.

'One thing,' said the Viking. 'When you're in Winnipeg, watch out for the widows. You're never too young.'

'OK.'

I slammed my door shut. The sound startled the geese. Their calls rose in pitch and volume, swelling to a metallic yammer. The flock lifted from the field as a single entity, 10,000 pairs of wings drumming the air, as if people were swatting the dust from rugs: white-phase and blue-phase geese jumbled together, tundra swans caught up in the confusion. We watched as the flock gained height, flashing when sun caught the backs and wings of white birds, then dispersing in straggling skeins and Vs, flying away to the north-west.

'My God!' the Viking exclaimed. 'Those birds!'

It was dark when we got to Winnipeg. The Viking dropped me at the hotel with the pink room. I thanked him; we shook hands. He warned me once more about the widows, wished me luck, and drove away.

6: MUSKEG EXPRESS

THE WOMAN IN the large blue plastic-framed glasses, navy fleece jacket, jeans and brand-new running shoes leaned forward, placed a paper cup of coffee on the burnished marble floor, then sat up straight again, fluffing out her long, curling, pale brown hair with both hands. She was about forty, with a smattering of faint freckles and no perceptible corners to her shoulders, nothing for the straps of a tote bag to find adequate purchase on. A squat brown leather suitcase waited like a dog beside her feet. To her left, a short, almost spherical man, sixty-odd, in an olive anorak, neatly creased charcoal trousers and shiny black leather shoes, perched on the edge of another wood-slatted bench, his hands resting one on top of the other on the pommel of a stick – an orthopaedic stick, provided by a hospital, with a scuffed rubber hoof and apertures in its metal tubing like the fingerholes in a recorder or antique flute. He had thick, fleshy lips and the bulging, eager eyes of an infant. His hair was the same tarnished silver as the Viking's, only longer and straighter, smoothed back like a bird's conditioned feathers with wax or oil. His hair shone like his shoes (he sported badges of light at top and toe) and he sat with proprietorial confidence, as if he owned the entire building.

The woman checked her watch. It was just before ten o'clock on the evening of 20 April, and we were sitting in the waiting area of Winnipeg's Union Station.

Our train was announced: the *Hudson Bay*. The spherical man was first up, grabbing the handles of a battered leatherette overnight bag with his left hand, wielding the orthopaedic stick in his right, and walking briskly with the rolling gait of a goose towards the uniformed VIA ticket collector at the gate. On the platform, he threaded his arm through the bag handles, hooked the stick on the crook of his elbow, reached up for both sides of the door-opening and hauled himself aboard with a blithe 'Hup!' I followed the brown-haired lady, her shoulders so self-effacing that her arms seemed joined directly to the base of her neck under the navy fleece jacket.

We located our tiny roomettes in the accommodation car without speaking a word. Upper and lower roomettes, seven feet by three, were dovetailed along both sides of the car – cabinets for living, surfaces and upholstery painted or dyed jade-grey, furnishings and features packed in tight as if on a submarine: cushioned seat, miniature toilet, hanging hooks, reading light, a whirring fan encased in a grille, a pull-out bed unlocked by a lever marked *Release/Déclencheur*. Each roomette had a pile of folded white hand-towels, a tube of conical waxed-paper drinking-cups, and a wash unit that folded down from the wall – a stainless steel basin with concentric rings shimmering in its clean, unblemished concavity, and no plughole but a narrow maw just below its wallside brim through which water drained sensibly whenever the unit was stowed upright.

The train jolted forward, had second thoughts, then eased slowly out of Union Station. I pushed the lever marked *Release/Déclencheur* and slid the narrow bed out like a morgue drawer across the toilet and cushioned seat. It locked into place: the roomette was all bed. The brown-haired woman and the spherical man were installed in jade-grey roomettes across the carpeted aisle, all three of us snug in our cubbyholes like whelks in their conches. I turned off the lights, released the blind and lay back as the *Hudson Bay* trundled across the confluence of the Red and Assiniboine Rivers and through the district of St Boniface, a cross of red bulbs glaring like a beacon from the roof of a tall building, the train rocking gently from side to side, couplings clanking and groaning as the carriages crossed points, getting the hang of the gauge. I heard these clankings through the chassis rumble and the roomette's medley of high-frequency shakes: the clatter of the basin troubling its latch, the grille trembling on the reading light, the clitter of curtain hooks, the sliding door rattling on its castors. Once the city lights were behind us, the window was a panel of stars, and when the train swung eastwards a crescent moon hove into view, its curve matched precisely to the curve of the steel basin stowed in the wall, its full circle visible with earthshine. There was comfort in the surrounding, rhythmic sounds, the carriage's gentle rocking motion, the cordial sensation of moving forward to a fixed end – proceeding mile by mile to the breeding grounds of the snow geese, with small towns passing through the night like proper nouns whispered in the ear: Plumas, Glenella, McCreary and Laurier; Dauphin, Roblin, Togo and Kamsack; Veregin, Mikado, Sturgis, Endeavour, Reserve.

This was the wheat line. The idea of a railroad linking Winnipeg to Hudson Bay had been proposed as early as 1812. Such a track, it was suggested, would provide the wheat farmers of western Canada with a direct route to European markets, far cheaper and simpler than hauling grain by train to Thunder Bay, then shipping it via the Great Lakes and the Saint Lawrence River out into the Atlantic through Cabot Strait. Although the water of Hudson Bay would only be ice-free for three months of each year, the port of Churchill was as near to Liverpool as Montreal, and 1,000 miles closer to the wheat fields of Alberta and Saskatchewan. Churchill was also 2,000 miles closer to the seaports of northern Russia than any other port on North America's eastern seaboard. When the construction of the Hudson Bay Railway was announced in September 1886, the *Manitoba Daily Free Press* celebrated 'the dawn of a glorious day. With an ocean outlet that will bring the prairie steppes and grain fields of this vast country as near to the British markets as the farmers of Eastern Canada and the seaports of the United States, the grand obstacle of distance will be swept away by a single stroke and the full granaries of this part will be placed on the thresholds of the British market.'

The grand obstacle of distance was not swept away quite so easily. The Hudson Bay Railway would not be completed until 1929. But work began in earnest, roadbed material wheeled forward on plank tracks from a work train, with rails, ties, hoisting devices, spikes and fishplate clamps following in a second train. Hundreds of Austrian, Italian, Finnish, Polish, Russian and Galician navvies sought work at the End of Steel, some hiking for twenty-two days, 468 miles along the track from Winnipeg to The Pas, sleeping on the timber sleepers. Some found jobs as stationmen, raising the grade, using gravel, rocks and thick vegetation to bring the roadbed to an even level; some ballasted the line, bolstering the ties that anchored the rails; some built bridges, fixing steel spans to concrete substructures to bear the track across the Saskatchewan River, Limestone River, Weir River and Owl River, and twice across Nelson River, at Kettle Rapids and Manitou Rapids. They worked sixteen-hour days, dressed in denim overalls, woollen shirts, cumbersome boots, and tweed caps or floppy, low-grade fedoras. They worked through winter, forty degrees below, and through summer plagues of mosquito and blackfly, with liquor forbidden, and no women allowed past Mile 412, except for nurses, who were taken in when needed. Their only luxuries were the Copenhagen snuff known as 'snoose' and the books, magazines and gramophone records provided by the Reading Camp Association, which also held classes in reading, writing, spelling, algebra and Canadian history.

After The Pas, the navvies had to deal with muskeg (an

Algonquian word meaning peat bog) and also with permafrost, the permanently frozen ground beneath the muskeg. It was essential to keep a layer of muskeg moss between the tracks and the permafrost, because any contact between the ice and the grade gravel would cause the ice to melt, resulting in a surface depression or gulch, and a sagging in the lines known as a sinkhole. Engineers sank pipes called thermosyphons several metres down into the ground to conduct heat away from the permafrost that supported the track, and the navvies erected ungainly tripods to carry telegraph wires alongside the line, because poles planted in the muskeg were casually ejected by the frost heave.

In some places the permafrost ran to a depth of 200 feet before it met the bedrock. One night, in the rain, a construction train working at the north end of the grade jumped the track and ploughed into the muskeg, dragging a string of eight flatcars behind it. The crew escaped unhurt, saw the train lying on its side, half-submerged in bog, and then walked to the nearest camp to send a telegraph message requesting the urgent dispatch of a repair train. When the wrecking train arrived, the crew accompanied it to the scene of the accident. But the locomotive and flatcars had disappeared. The crew stood dumbfounded. The derailed train had ploughed right down to the permafrost and proved such an excellent conductor of heat that it had melted the ice and sunk on its own warmth until completely swallowed by the ground. This was land that could eat things up.

The track reached Churchill on 29 March, 1929. The last spike, wrapped in tinfoil ripped from a packet of tobacco, was hammered in to mark completion of the project: an iron spike in silver ceremonial trappings. The first shipment of grain followed later that year. A ton of prairie wheat, grown in southern Manitoba, packed in 1,000 two-pound canvas bags for distribution in England, was shipped by James Richardson

& Sons Ltd on behalf of the Hudson's Bay Company in the SS *Ungava*. In 1932, ten ships crossed the Atlantic to load wheat in Churchill. The tenth, the *Bright Fan*, with 253,000 bushels of wheat on board, struck an iceberg in Hudson Strait and sank. This disaster was a foretaste of decline. Shipments of wheat from Churchill have slumped since their peak in 1977. The Hudson Bay Railway runs at a loss and is heavily subsidized. There are frequent calls for the closure of the port.

From Winnipeg the line ran west to Portage-la-Prairie, then turned northwards for The Pas, with Lake Manitoba and Lake Winnipegosis to the east, and to the west, outside my roomette's window, the dense spruce forests of Riding Mountain, Duck Mountain and the Porcupine Hills. Lying back in the dark, I imagined snow geese passing overhead towards Churchill, coterminous with millions of bushels of prairie wheat.

The Hudson Bay Railway wouldn't exist without wheat. It ventures into the hinterland of northern Canada, a region known to early mapmakers as the Barren Grounds. Even on a modern map the line seems a line of enquiry, a speculative filament or strand, adrift in the wide open. Thirty-six hours away, 1,000 miles to the north-east, a vast grain elevator waited at the mouth of the Churchill River, with holding room for 5 million bushels of wheat and the capacity to deliver 60,000 bushels an hour to ships moored in the deep-sea berths. But first the *Hudson Bay* had to make its way through boreal forest and across tundra, proceeding with due caution to negotiate sinkholes where the track had dipped or 'wowed' on the thawing of the ice below.

*

I WOKE TO FORESTS OF spruce and aspen, and a mallard drake, a greenhead, flying alongside the train, just outside my window, racing me northwards. I got dressed and walked through to the dining-car. The short, almost spherical man with the slicked-back hair had already eaten and had turned his chair to face the window across the aisle. Wearing neatly creased charcoal trousers, shiny black shoes, and a white short-sleeved button-up shirt with a plain white T-shirt as a vest beneath it, he was watching innumerable spruce trees drift by, the deep green of the conifers broken here and there by slender white aspen trunks. His hands were resting on the pommel of the orthopaedic stick, and he was leaning forward a little, trusting his weight to it, as an old wizard, a Merlin or Gandalf, would trust his weight to a magic staff.

'So?' he said. 'How do you like the Muskeg Express?'

A deep, husky, croaking voice, a bullfrog's voice, the s sounds of Muskeg Express gathering a wet, lispy sibilance as they passed through his mouth.

'It's not exactly express.'

'First time, right?'

'Right. You?'

'Nineteenth.'

'Nineteenth?'

'Yep. Don't even live in Churchill.'

'Are you going to visit somebody?'

'No. Just here for the ride. Just taking the train.'

His brow was agile, mischievous. He didn't look at me; he kept his eyes on the trees; he looked up when the woman with the soft, airy brown hair came into the dining-car, wearing her large blue plastic-framed glasses, jeans, bright white running shoes, and a white T-shirt with an abstract grey motif on the front of it, hanging baggily off her sloping shoulders.

'Morning!' he said.

'Morning,' she said.

'Morning,' I said.

'Morning,' she said, sitting down opposite me, at the only place laid for breakfast. Behind the blue plastic frames a haze of browns: light, tumbling curls of brown hair, hazel eyes, pale brown eyebrows, the smattering of freckles. We ate breakfast together while the man with the metal stick and the bullfrog's voice gazed out of the window at the trees. The woman's name was Brenda. She lived in Winnipeg and was visiting her mother in Churchill. She worked in a bank; her husband sold garden machinery.

'Oh, Mark's a born salesman,' she said. 'He has the gift of the gab. He could sell you the shirt off your own back. He'd sell you a mowing machine if you didn't even have a lawn to put it on.'

'How long are you going to spend in Churchill?' I asked.

'Just a couple of days. Any longer than that, I start to miss my dogs.'

'What kind of dogs do you have?'

'Dobies.'

'Dobies?'

'Dobermann pinschers? I'm a breeder? I've got fourteen Dobies.'

The abstract grey motif on her T-shirt resolved itself into three head-and-shoulders portraits of Dobies. Brenda pulled a leather wallet from her jeans, the wallet fat not with notes but with photographs of Dobies, sleek on podiums, Brenda standing beside them, brandishing silver trophies, the camera flash flashing off the trophies and her blue-framed spectacles.

'This is one of my girls,' she said. 'And this is my eldest boy.'

'Good-looking dogs.'

'Oh, I love my dogs.'

The rattling metre of carriages on rails. Sometimes the forest opened out on frozen lakes, reaches of crisp white scurf fringed with dark spruce trees, the train following the curves of the lakes with a loose and sensual flexure. On both sides of the track I could see signs of the thaw: the year's first pools and streams in the muskeg, and Canada geese, herring gulls, mallard and lesser scaup sitting on patches of open water, the first birds up from the south, a season's vanguard. Settlements held to the lakesides – clusters of ramshackle shacks, prefabricated cabins, satellite dishes, skidoos, Dodge Ram and Chevrolet S-10 pickups, and boats upturned at the ice-edge, their hulls bared like haunches to the keen cold. Sometimes the *Hudson Bay* entered districts of forest ravaged by wildfire, a constant hazard in summer, when the duff layer of dry needles and feathermoss on the forest floor was primed to kindle and smoulder if lightning gave it half a chance. The blazes left the spruce trees with a scraggy fletching of short branches and green needles at their tops, or strimmed them bare like cabers: charred poles footed in the ash.

'Are you on vacation?' Brenda asked me.

'Not exactly. I'm watching birds.'

'OK,' she said, nodding.

'Snow geese. They migrate up to Hudson Bay from around Winnipeg. Some of them nest very close to Churchill.'

'OK. I think I've seen them.'

'Snow geese?' said the man with the stick.

'Yes,' I said. 'I've been following them up from Texas.'

'You like trains?'

'Yes.'

Brenda pushed back her chair and stood up, wiping her hands with the paper napkin.

'I'm just going back to my roomette,' she said. 'I'm going to rest up back there.'

'Don't be late for lunch!' the man said.

'Because they're going to run out, right?' Brenda replied, smiling, before disappearing through the door into the accommodation car.

Across the aisle, the man was still leaning forward, in earnest, hands on the pommel of the orthopaedic stick, facing the window. Spruce trees drifted past, with the odd duck or gull or Canada goose, the white startlings of iced-over lakes.

'How long are you going to stay in Churchill?' I asked.

'Just tomorrow,' he said. 'Quick turnaround.' He kept his eyes on the window.

'Do you live in Winnipeg?'

'Live in Thunder Bay. You ever been to Thunder Bay?'

'No.'

'It's on Lake Superior. That's one of the Great Lakes. Know the Great Lakes?'

'No.'

'Huron, Ontario, Michigan, Erie, Superior. Spells *HOMES*. That's how you remember.'

The s sounds of 'spells homes' delivered with a raspy hiss.

'Are you going to visit someone in Churchill?' I asked.

'Nope.'

'Is this a vacation?'

'Guess so. I like trains.'

The man's name was Marshall. He *loved* trains. After breakfast, in the dining-car, with the boreal spruce forests and frozen lakes of central Manitoba filling the windows on both sides, he told stories about the railways, holding forth in his deep, croaking voice. He seemed to have been waiting for an audience. His reminiscences gathered momentum; he banged the stick's rubber hoof on the floor in front of his shiny black shoes for emphasis, his brow pulsing like a squirrel, his eyes trained on the window as he talked, the whole carriage lit as if for a photograph when the train met a lake and sunlight flashed off the open snow surface. Marshall had loved the railroads since 1947, when, aged fifteen, he'd run away from home in Montreal and boarded the first train he could get out of the city.

'I don't know if it was stubbornness, pride, sense of adventure, independence or somesuch,' he said, 'but that was the last ticket I bought for a long, long time. Jumped my first rattler, which is what you'd call a freight train, at the Richmond watertank outside Toronto, and rode in a boxcar to Washago with an old hobo named Jerry. We stayed overnight in the derelict station they had there, and when we woke up at daylight the sun peeked in at an angle into the ticket office which was all dust and cobwebs, so I got up to pull down the shades to keep the sun out and twenty-one dollars fell out of the blind in single dollar bills. A bonanza! Then me and Jerry jumped a rattler to Parry Sound where we went straight to a beanery run by Bert O'Dell for breakfast, lunch and dinner all rolled into one, and after that I jumped a rattler to Capreol, and the conductor in the caboose at the tail of the train was Fred "White Pine" Thomson, he was six foot six inches tall and white as a glass of milk, and another conductor was Boob Graham, named "Boob" on account of his wife wearing the pants, and then I wound up in Sioux Lookout and got myself recruited for forest fire duty. We all slept in tents and I slept next to a regular wino named Shotgun Benny Ferguson and we worked from sunrise till sunset till we were played out, burnt out, smoked out, dirtied up with smoke and soot and ravaged by came-from-hell mosquito and blackfly. Never wash, that was the thing: keeps the mosquitoes away and your in-laws also.'

Marshall had lived on the railroads, sleeping in boxcars, meeting old friends in the railyard communities called hobo jungles, jumping rattlers from one end of Canada to another, Atlantic to Pacific shore: the Rocky Mountains, the St Lawrence River, the granite shield of Ontario, the open plains of Alberta and Saskatchewan. Now he was holding court in the dining-car, sometimes lifting one hand from the pommel of the stick and gesturing, or banging its hoof hard on the floor as the *Hudson*

Bay rocked and trundled on north-east through Orok, Finger, Budd and Dyce, his stories rehearsed and fluent, annals of a bard or troubadour.

As a Canadian and, in theory, a taxpayer, Marshall had considered himself one of the owners of the railway. Why should he pay twice for the same service? When a constable had tried to arrest him on a train in Biggar, Saskatchewan, Marshall had given him a piece of his mind. 'We all of us pay income tax,' he said, his rasping voice gathering force as he mimicked his own younger self, squirrel-brow working busily above bulging, gelid eyes, 'and this son-of-a-whore railroad runs at a deficit of 100 million dollars a year of taxpayers' money! Technically speaking, we're all shareholders in the company! I know that between Halifax and Prince Rupert there's one spike with my name on it, and all I'm doing, mister, is looking for the damn thing!' Marshall paused and looked straight at me for a moment, as if I were the constable and Marshall were the teenager waiting for some rejoinder to his tirade. But I said nothing. Marshall turned back to the window, still flickering with spruce trees, and gazed out absently, in a fugue of his own devising.

'What kind of people were jumping the trains in those days?' I asked.

'Oh,' said Marshall, snapping back, 'these were folk like Suitcase Simpson, he never had a backpack, and Boxcar Kelly, I guess he was born on a boxcar, he'd been on them most of his life, and Branchline Shorty, always on a branchline, never on a mainline, and Hollywood Slim, who wanted to be a movie star, he'd been in Hollywood in the twenties and thirties, he wanted to be in those Griffith D.W. pictures but he never made it, and Slim's girlfriend Gravel Gertie, and Tommy Slater from Sapperton, eighty years of age, used to work in Boston lighting gas lamps, got laid off when electrics arrived, hadn't worked since, and Tommy's girlfriend Nightmare Alice, she *was* a nightmare –

tough as nails, swore like a trooper, drank with the best of the guys and fought like a bunch of wildcats – and Sticks Wilson, walked with a crutch but if he was chasing a boxcar he threw the damn thing in and ran like an athlete! Sometimes you'd travel together and sometimes you'd catch up with them at the jungles which usually you'd find on the outskirts of railroad marshalling yards, places like Kamloops, Boston Bar, North Bend. We stole potatoes, stole corncobs and carrots. If you had money you spent it on wine: Calona Royal Red at three dollars seventy-five a gallon – drink enough of that, you'd start singing "O Canada" in Chinese!'

Marshall thought of the trains as his education. 'I went to university,' he said. 'Sure I did. Majored in bum-ology and tramp-ology, coast to coast on tea and toast!' In the 1930s, Marshall explained, the singer Hank Snow had had a hit called 'I'm Moving On', and the hoboes liked to say, 'I'm going to pull a Hank Snow', which meant that they were going to jump a train and move on. Marshall talked about greenhorns – boys who thought themselves hoboes but knew nothing of the world or the railroads in it – and about freezing winters during which greenhorns and hoboes alike kept to the cities, staying in Salvation Army hostels known as Sally-Anns. When a constable asked them where they were going, they'd say, 'My sister's place!' and the constable would ask, 'Who the hell's your sister?' and they'd reply, 'Sally-Ann!' There was a law against vagrancy – the penalty was a fifty-dollar fine or thirty days in the can, but this would be waived if you left town within twenty-four hours, in which case you'd get a suspended sentence called a floater. Marshall was always on the lookout for the Canadian National Railway security men, the railroad bulls. Whenever he was arrested and asked for his address, he'd say, 'CN19586. It's the last boxcar out at Transcona Yard. You can't miss it!' Somehow he wound up at Maple Creek, Saskatchewan, where he operated a small bulldozer as part of a team laying a pipeline, but he got restless and hitched a ride out of town without paying his hotel bill. He was arrested in Regina and hauled up before a judge with a notorious drunkard named Haywire Walker. Walker was a squaw man, which meant he had a taste for Indian women, and he'd been accused of selling liquor to Indians. Winter was fast approaching, and Walker had his eye on some comfortable accommodation. When the judge found him guilty and sentenced him to thirty days, Walker said, 'Is that all? It's getting cold!' The judge said, 'You're a wise guy, huh?' and raised the sentence to sixty days. Marshall was sentenced to thirty days in the Regina bucket for skipping the hotel in Maple Creek, and when the month was up he heard they were looking for a cook, so he said he was a chef with wide experience and got the job as gaol cook. He went down to the cells with pork chops taped to his calves with sticking plaster and exchanged the pork chops for the prisoners' tobacco allowances.

'Did you actually know how to cook?' I asked.

'Couldn't parboil horseshit for a Hudson Bay trapper,' Marshall replied. 'Not if my life depended on it!' He banged the floor three times with his stick, as if to summon an equerry, then broke into a complex, virtuoso laugh that began life as a wild cackle, slowed down and deepened to the splutter of an old boat's engine starting, then thinned out into a nasty asthmatic wheeze, as if his lungs had turned to thistles. Marshall, holding a white handkerchief to his mouth, cleared the wheeze with a volley of rough, globby coughs and spat whatever had arrived in his mouth into the handkerchief, which he then held out at arm's length and eyed intently, inspecting his pulmonary handiwork.

'Time for a rest,' he said, returning the handkerchief to the pocket of his creased charcoal trousers. He stood up and walked towards the end of the car, his rolling gait accentuated by the rocking motion of the train. A leather belt, like the ribbon round a gift plum pudding, encircled Marshall at his widest point – an equatorial band that marked out his northern and southern hemispheres. He turned at the carriage door.

'Don't be late for lunch!' he said, and disappeared, my head ringing with the singsong of boxcar, bogie-truck, gondola, caboose.

Wekusko, Button, Dunlop, Medard. Back in the tiny jade-grey roomette, I gazed out at the parade of spruce trees, their sharp nibs like the teeth of a saw, notching sky. The rattling four-syllable figure of the wheels crossing the rail-sections made a bass continuo above which the roomette's several percussive lines (the clitter of curtain hooks, the trembling of the grille on the fan, the shake of the stowed basin on its latch) played like maracas, timbrels, castanets. When I walked back along the train to the dining-car, I found Marshall sitting alone at his table, eating a sandwich. He was holding one half of the sandwich in each hand and taking alternate bites: left, right, left, right. Two mats were laid at the table across the aisle where Brenda and I had eaten breakfast, and when Brenda came into the dining-car we sat in these appointed places, leaving Marshall with a table to himself. After lunch the chef tuned the radio in the galley to a live broadcast of a Cree pow-wow, voices wailing pell-mell in the static hiss and crackle. Brenda pushed her chair back and stood up, wiping her mouth with a napkin.

'Don't worry,' she said to Marshall, 'I won't be late for dinner.'

'I didn't say anything!' he protested. 'Who heard me say anything?' The note of a child's pleading was incongruous in his gruff, bullfrog's voice. Brenda left the dining-car.

'Did I say anything?' Marshall asked me, grinning, raising his hands, palms upwards, a gesture of blamelessness.

'I'd love to hear more about the trains,' I said.

'Ah!' said Marshall. 'Step into my office!' He indicated the empty chair opposite him. I crossed the aisle and sat just where he had bidden. The orthopaedic stick lay across the white tablecloth close to the window, next to a smoky white plastic pill-tidy with seven compartments marked Mo Tu We Th Fr Sa Su in black letters, each containing a salvo of five or six tablets and capsules. Marshall took a deep breath and began reeling off clauses of the Railroad Act, which he seemed to know by heart, and then he skipped to the signalling procedures used on each of the different branchlines of the Canadian railway network, and suddenly it was 1960 and we were in the Yukon, where Marshall was cooking for crews working on the White Pass Railroad, and then he was cooking for geological survey teams investigating the furthest reaches of Alaska and Labrador.

'Slow down!' I said.

'Oh sure!' Marshall replied. 'I can talk! People tell me I never shut up! Thing is, in 1940, when I was in school, they'd run out of vaccination needles due to the metal shortages caused by the war, and all they had left were gramophone needles, and when it was my turn for vaccinations they jabbed me with a gramophone needle and I've never shut up since. Must be in the blood!'

He laughed, began to wheeze, produced the white handkerchief from his charcoal trousers, held it to his mouth and cleared his lungs with a volley of coughs, his eyes watering and bloodshot. He breathed deeply for a moment, mustering strength, then went on with his stories as if nothing had happened, talking as the *Hudson Bay* made progress north-east through Lyddal, Odhil, Hockin and Sipiwesk and took on passengers at Thompson – Cree families from the reservations at Nelson House and Split Lake, wearing baseball caps, smoking cigarettes, crowding the dining-car tables behind us and embarking on epic poker games, the broadcast pow-wow soon drowned out by shouted wagers and disputes and the slap-slams of one-dollar and two-dollar coins palmed hard on the Formica tables.

'I grew up in Montreal,' Marshall continued through the hubbub, eyes bulging, burly forearms resting flat on the white tablecloth, 'in a neighbourhood of White Russian Jews, garment workers mostly: St Viatur, Fairmont, Clark Street, St Lawrence, the kind of places the book writer Mordecai Richler writes about in his book *Horseman of Saint Urbain*. We were six children. My parents split when I was twelve. I went to foster parents and ran away like I told you. Hit the sawdust trail. Grabbed an armful of rattlers and headed for the sunset. Most of the people travelling the railroads were seeking one thing or another, coming from broken homes or looking for love or a better way of life or a pot of gold. It was a way of escape. Nowadays they use drugs, but we used a cheap bottle of wine and a boxcar. Faith was the thing. You had to have it: faith. There was a Sally-Ann in Vancouver called Harbour Lights where they had a prayer meeting every Sunday evening, and the Sally-Ann in Edmonton was run by Captain Lesley. They had these nice glass panels with patterns etched in the glass, and when a hobo got refused a meal he'd throw a brick through the glass panels, so they soon forgot about the glass and bricked it over. That's where I met Cartier the Booster. He stole from department stores. He'd ask me my measurements and I'd say, "Twenty-eight-inch waist, fourteen-and-a-half-inch collar", and in fifteen minutes he'd come back with a whole new outfit from the Army and Navy – twenty dollars' worth of clothes and he'd sell them to me for all of five. In Regina there was Crooked Nick Scotty Woods with his colostomy and vanilla extract, the vanilla had a small alcohol content so Crooked Nick Scotty Woods drank it morning, noon and night and smelled like a walking bakery as a consequence. The Can Heat Artists drank all the methylated spirits they could squeeze out of Sterno stoves, and the Baysie Boys drank bay rum hair tonic, so much of it that some of the Chinese grocery stores bought bay rum in forty-five-gallon drums and sold it off in two-ounce bottles named spark plugs, and if you brought back an empty spark plug bottle they'd exchange it for a cigarette. Then there were the skid row bars – the Occidental Hotel in Winnipeg, corner of Main and Logan, ten cents for a glass of beer, the Dodson, the Travellers' in Vancouver, the Royal in Edmonton, 96th Street . . .'

Marshall drew breath. There was a roar from the gamblers: a full house, a flush, a daring bluff, a scam. Thick grey swirls of cigarette smoke hung over the baseball caps.

'You had a twenty-eight-inch waist in those days?' I asked.

'That's right. Twenty-eight-inch waist, weighed ninety-eight pounds. Fifty-one years later it's a fifty-four-inch waist and two hundred and forty-six pounds. When the good Lord created me, He just didn't want to quit. But no cholesterol! Not a drop! As far as I'm concerned, Cholesterol's a little mining town in Nova Scotia. My brother gets me this wild garlic. Once a year he goes hunting in the Laurentian Mountains in Quebec for some of their wild garlic, and once a year I get a package of wild garlic in the mail. I pickle it in olive oil, keep it in the fridge, take one clove every morning for a month, wash it down with a glass of orange juice. When I go to my doctor for a check-up he can't believe it. "It's amazing!" he says. "This man has no cholesterol!" I tell him about the garlic, and the doctor listens hard, because here's a man weighs close to two hundred and fifty pounds and he's got the cholesterol of a baby girl! "The garlic!" I tell him. "It's all in the garlic!"'

A rowdy cheer went up from the tables behind Marshall, as if to celebrate the miracle of garlic. Cree families were crowding the games, jostling for a sight of the cards, wearing quilted lumberjack jackets, bandying strategies, gambits, accusations, critiques. Marshall turned to look behind him, then faced me again, chuckling.

'OK.' he said. 'That's it. Time's up.'

He rubbed his eyes like a tired choirboy, then pushed his chair back and heaved himself up, pressing down on the table-edge. He slipped the plastic pill-tidy into his trouser pocket, grabbed the metal stick, and set off down the aisle between the tables. I followed him back through to the accommodation car. Marshall had to turn sideways to ease his girth through the narrow intercarriage doors. He paused at the entrance to his roomette.

'Don't forget!' he said.

'What?'

'What do you think?'

'Garlic?'

'No!'

'I don't know.'

'Dinner!'

Marshall whipped back the door-curtain and stepped inside, wheezing.

I sat back in my roomette. The train trundled on, wary of the heave and wow of muskeg and permafrost under the roadbed, and places kept passing like proper nouns murmured close to the ear: Pikwitonei, Wilde, Arnot, Boyd. The spruce trees began to thin out as we drew closer to the timberline, the transitional area or ecotone that separates boreal forest from open tundra. A bald eagle spiralled on a thermal; a few mallard, redhead and lesser scaup rested on patches of open water; pairs of Canada geese flew alongside the train, outside my window, white chinstraps gleaming in the sun, their long black necks stretched out like arms, with bills for index fingers, pointing north-east towards Hudson Bay. Each time I saw these birds I thought of snow geese. I checked for tell-tale shapes above the jagged spruce line: Vs, Ws, echelons, skeins. We crossed the Nelson River at Manitou Rapids and I looked down through the bridge spars at water surging in the narrow channel, fraught with eddies and spasms, such a volume of melted ice it seemed the gorge was the node of the thaw, with winter draining through it to the sea.

It was almost two months since I'd flown from London to Houston. I remembered how keenly I had anticipated my first sight of snow geese, glancing up at the stuffed bird above Ken in the Sportsman's Restaurant at Eagle Lake, then waiting on the prairies by the blue Cavalier at sunset, listening for the faint calls of returning flocks, the sound of halliards flicking on yacht masts, or terriers yapping. I remembered the flocks Jean had pointed out from the Greyhound, working north to Minneapolis. I pictured the wild, drumming blizzards of blue-phase and white-phase snows wheeling above Sand Lake, the waving strands of geese that had passed over Riding Mountain as fluff from cattail seedheads skiffed past the Viking on a spring breeze. I wondered if I had seen any of these birds more than once; if our progress north had been coincident.

Marshall ate dinner alone at his table on the right-hand side of the train. Brenda and I sat at our usual places across the aisle, the dining-car full of smoke and poker sounds. We retired to our roomettes for the second night. I pressed the lever marked *Release/Déclencheur*, pulled the bed from beneath the mirror, turned off the lights and let the blind up as far as it would go. The sky was clear; the window framed an abundance of western stars: Orion, Gemini, Auriga, Perseus, Cassiopeia. The bed was warm, soft and snug, underbraced by the rattling four-syllable mantra of the rails, rocking gently as the *Hudson Bay* plied the uneven taiga grade. I kept hearing Marshall's husky, croaking voice, textured with wheeze, the muffled beats as the stick's pale grey scuffed hoof hit the floor in emphasis. Hoboes, rattlers, greenhorns, Sally-Anns. To the north, I could see Ursa Minor, the Little Bear or Dipper, with Polaris, the North Star, at the tip of its handle. *Ursus* is Latin for 'bear'; 'Arctic' comes from the Greek word *arktos*, meaning 'bear': the Arctic was that precinct of the globe that lay beneath the jurisdiction of Ursa Major and Ursa Minor.

I lay awake, thinking of home. Not just of the ironstone house – my mother's evening viola scales coming up the stairs – but also of the London flat in which I'd been living, the streets around it, the faces and voices of friends, the things we laughed about. Such images had occupied my mind with increasing frequency ever since my stay in the white motel room in Aberdeen. In that room my curiosity, my appetite for the new, seemed to tire or slacken, perhaps because I was lonely, or because I felt for the first time that my journey north with the snow geese was not quite the shout of freedom I had presupposed. I was aware of another impulse that, if not the opposite of curiosity, was certainly resistant to the new or strange and sympathetic to everything I could remember and understand. This wasn't the acute longing I remembered from hospital, that desperate nostalgic desire to return to the circumstances of childhood. Lying awake on the train, what I felt was no more than a mild ache, bittersweet, an awareness of separation from things I loved, an almost corporeal inclination towards familiar ground. It was as if I existed between two poles, the known and the new, and found myself drawn alternately from one to the other.

For James Copland, whose *Dictionary of Practical Medicine* was published in 1858, nostalgia was a cause of disease rather than a disease itself. But Copland, drawing extensively on both Hofer and Larrey, took homesickness extremely seriously. 'The suggestions of memory,' he wrote, 'in continually haunting the mind of him who has removed, for the first time, from the scenes of varied enjoyments and strong excitements, to places remote, not only from these but from all other attachments, particularly if he be doomed to different avocations from those to which he had become accustomed, are amongst the most

distressing of the numerous ills that embitter the destiny of man.'
Copland reiterated the old belief that mountain-dwellers were
particularly susceptible to nostalgia: 'Numerous examples of the
effects of continued longing for the scenes of early life occur
to the medical practitioner; but they are most common amongst
the natives of the high lands, as those of Switzerland and of
Scotland, when they migrate to the low countries, where this
feeling is heightened by the influence of a more depressing air
upon constitutions formed in the pure and cold atmosphere of
more elevated regions.'

Copland characterized the first signs of nostalgia as 'unusual
reserve, sadness, distaste of amusement and of occupation, a
continual recurrence to the various circumstances connected with
home, and expression of regret at removal, with a desire of
returning and of enjoying those pleasures which the imagination
is constantly presenting in more glowing colours than are real'.
Decline, pallor, emaciation and 'painful rumination' may follow.
'The patient nurses his misery, augments it until it destroys his
nightly repose and his daily peace, and ultimately devours, with
more or less rapidity, his vital organs.'

How should physicians endeavour to treat such attacks of
return-suffering? 'Nostalgia,' Copland remarked, 'requires more
of moral than of medical intervention.' He suggested that 'kindness, encouragement, and exciting hopes of soon revisiting the
scenes for which the patient longs, are generally of the greatest
service'. Hofer had described the case of a young man from Bern
who had left home to study in Basel. After a period of dejection,
the student developed anxiety, palpitations and a continuous
fever – symptoms that grew so severe that death appeared
imminent. Prayers were recited on the student's behalf. The
apothecary, who had been summoned to deliver an enema,
observed that the patient was suffering from homesickness and
advised that he be sent home in a litter immediately. As soon as

the student heard this and saw that preparations for his return were being made, he drew breath and became calmer, and after several miles on the road from Basel his complaints had abated significantly. Before arriving in Bern, he had already recovered. 'What strikes one most in the sparse literature on help for the homesick,' write the authors of a 1996 review, 'is that often only returning to the old home environment brings real relief.'

And there *was* relief, even as we left the hospital for home, after dark, in the new year: a feeling of inevitable movement, as if I were attached by lines of elastic to the ironstone house, elastic bungee cords that gave an illusion of freedom but exerted a pull, a return-pressure, as soon as they were stretched and tested. There was comfort even in the process of going back, seeing the lights of houses arranged in patterns as fixed as constellations, the road dipping or curving exactly as you remembered and expected, the shapes of signs, buildings, trees and spires appearing on cue, each one its own fulfilment. We pulled up at the house, and my mother and father helped me to the front door, over the stone floor of the hall, up the stairs, the landing floorboards creaking correctly underfoot, into the dressing-room, the curtains drawn, the rightness of things in their allotted places.

The *Hudson Bay* made slow, fitful progress through the night: Nonsuch, Wivenhoe, Bird and Charlesbois; Herchmer, Kellet, O'Day and Back. I drifted in and out of sleep, constellations wheeling across the blue-black window on the hub of the North Star. Shooting stars, particles burning up as they entered the Earth's atmosphere, streaked with lazy, casual brilliance, some moving so slowly you could track the transverse paths and knock off a wish while the glow still hung there. In the early morning a band of faint greenish white light appeared in the north, a shallow curve, with vertical bars of light rolling through it like billows through a bolt of silk, and for a moment I wondered if I were dreaming these northern lights. I'd read about spectacular nightshows – bold light streamers, clouds of red and blue fluorescence lurching from one side of the sky to another, accompanied by loud hisses and cracklings – but this was a modest display, a hanging soft-lit cirrus drape, swaying and undulating as if a wind disturbed it: silk sheers luffing in an open summer window. I felt the imminence of the far north. The breeding grounds of the snow geese were close now: thousands of geese would soon be settling to nest at La Pérouse Bay, just thirty miles east of Churchill.

The term 'aurora borealis' was introduced by the French astronomer Pierre Gassendi in 1621, from the Latin *aurora* ('dawn') and *borealis* ('north'). Boreas was the god of the north wind in Greek mythology; the original scientific name for a snow goose was *Anser hyperborea*: 'goose from beyond the north wind'. Some thought the northern lights were old women dancing in white gloves, gods dancing, flames glimpsed through cracks in the firmament, the fire of volcanoes erupting close to the pole. Others understood them to be reflections off the shields of Valkyries, off vast flocks of greylag geese, off the turbulence of threshing whales or the wings of swans trapped in polar ice. They were spirits playing football with a walrus skull, the arc of a burning sky-bridge by which gods could pass from heaven to Earth, the glow given off by huge herring runs and other fish schools in northern seas, or the light glancing off a fox with glittering fur as it ran across the mountains of Lapland. Or they were caused by particles streaming from coronal holes in the surface of the sun – ionized hydrogen protons and electrons, and smaller quantities of helium, oxygen and other elements, discharged at velocities sufficient to escape solar gravity, travelling through space as the solar wind, entering the Earth's atmosphere at the magnetic poles and colliding with atmospheric atoms and molecules, which absorbed energy from the impacts and emitted energy (including infrared, ultraviolet and visible light) as they settled back to their neutral state. I lay next to the window, watching the aurora, the faint light rippling like the hem of a gown. The curtain hooks clittered; the sliding door rattled on its castors; the stowed basin shook on the latch.

I slept for three or four hours. Marshall woke me, rapping on the door.

'Get up!' he wheezed. 'There are caribou!'

I got dressed. Marshall and Brenda were in the dining-car, standing in the aisle, looking through the left-side windows. The *Hudson Bay* was travelling through tundra now. The land seemed laid open to the sky, flayed, with no tree cover but for the occasional haggard, stunted spruce. The tundra surface was rough, almost corrugated, blotched with greens, browns and sere yellows, finished with bristling sedge tussocks and glistening snow patches. Twelve Canada geese were flying alongside the dining-car in a short V, keeping pace, honking with ribald vigour. A few cumulus clouds floated like dirigibles above the clean horizon line.

'Look!' commanded Marshall, jabbing a stubby finger at the tundra.

I looked. A herd of caribou, fur tones well matched to the tundra's sere browns, were cantering away from the train, towards the white dirigibles. A small contingent of the Kaminuriak herd of barren-ground caribou, these reindeer were themselves migrants, wintering in the taiga forest, trekking northwards across the tundra towards calving grounds each spring. Their movements, like those of snow geese, free-tailed bats and common swifts, were tied to Earth's yearly swing round the sun. Migration was their way of dealing with the tilt.

We took our seats, Brenda and I opposite each other, Marshall at his usual table across the aisle.

'I saw the northern lights,' I told Brenda. When she turned towards the window I could see tundra moving across the lenses of her blue-framed glasses.

'That's good,' she said, smiling, and nodding, as if agreeing that I had indeed seen the northern lights.

'Aurora borealis!' Marshall declared, without explanation or commentary, as if he were a commissionaire, announcing the aurora's arrival at a grand ball. He was looking away from us at the tundra outside the right-hand, eastward window.

'I'd never seen the lights before,' I said.

'Yep. Aurora borealis,' he repeated, still gazing at the flat barren grounds. 'The northern lights.'

'I guess we're pretty close to Churchill,' Brenda said.

'Ten minutes,' Marshall said.

It was already past eight o'clock. The *Hudson Bay* was more than an hour behind schedule. Brenda fluffed out her brown hair.

'You packed?' Marshall demanded.

'Not yet,' I said.

'You should be.' He checked his watch. 'Five minutes.' Marshall's leatherette travel bag was ready at his feet; his olive anorak was draped over a chair.

I followed Brenda back to the roomettes and packed my things. The *Hudson Bay* pulled into Churchill exactly as Marshall had predicted. The three of us stepped down to the platform. We followed the baseball caps and lumberjack shirts of the Cree families through the small terminal building. The grain elevator loomed at the end of the line, on the edge of Hudson Bay: grey, massive, brutal, conceived on a different scale to every surrounding structure. At the terminal the track frayed into six or seven spurs along which cars of grain could travel the last few hundred yards to the silos. Black-and-white snow buntings flitted from spur to spur, pecking at spilled wheat. A pair of Canada geese honked in the muskeg between the station and the frozen Churchill River. The cold was bracing. The light was silver and fierce. The sense of space was dizzying. I scanned the sky, looking for snow geese, but there was only the elevator, rearing at the end of the line like a vast tomb. An emblematic design: the railroad, time; the elevator, doom.

I shook hands with Brenda and Marshall. We were wearing gloves and our handshakes were clumsy. Marshall set off for Gypsy's Bakery, where he intended to wait out the day before boarding the train again that evening. He carried the leatherette bag in his left hand, the orthopaedic stick in his right, and he walked away down Kelsey Boulevard with the brisk, rolling gait I had first seen in Union Station, Winnipeg, just thirty-six hours before. I watched him go, then lugged my bags to the bed and breakfast on Robie Street.

7: CHURCHILL

IT TOOK ME BY SURPRISE, that I should feel so lost without geese. For several days it seemed I did nothing but walk, tramping from one end of Churchill to the other, establishing the beginnings of a home range, getting to know the axis of Kelsey Boulevard and the streets that led off it eastwards to La Verendrye Avenue, orientating myself first by prominent landmarks like the grain elevator, the Anglican church and the Northern Stores, and then by smaller points of reference: a pink garage door, a blue roof. It was the end of April, but on the front of one house silver tinsel and a string of gaudy lights still picked out the outline of a Christmas tree, and in a yard a handmade sign, bordered with green tinsel, read simply, *Peace on Earth.*

There wasn't much snow. Snowmobiles and rusting Bombardier snow vans were pulled up on the dirt shoulders of the roads, out of their element. Four-wheeler all-terrain vehicles sped past, most of them ridden by children in quilted parkas, hoods thrown back, hoops of fur trim bouncing on their shoulders, and there were more ATVs parked in Hudson Square, where wire netting defined the back corner of a baseball diamond and parallel lines of snow were set hard in the tier-angles of the timber bleachers.

The elevator loomed at the end of the line: a central section for cleaning grain and an annexe on either side, like wings, for storage. I walked along the spurs towards it, the rails spattered with bunting guano, the elevator tomb-grey, derelict, its windows smashed in by frost. A long enclosed gantry of rusting corrugated iron led to a gallery eighty feet above the wharf, with red chutes that would swing out over the water when the ice cleared, grain schussing down them into ships' holds. A harbour tug, the *George Kydd*, rested in dry dock; two yellow dredging cranes sat on caterpillar tracks; steel masts carried coronas of arc lights high above the wharf.

The Churchill River was frozen, the ice buckled where it had driven into the quayside, crumpling on itself, heaving up boulders and sooty rubble. Behind a breakwater, ice clamped the hulls of a larger tug, the *Keewatin*, and the four grey barges that in summer would take fuel and machinery to Inuit communities further north: Arviat, Whale Cove, Rankin Inlet. Men in hard hats and coveralls were working on the barges, preparing them for the season. A welding torch fired deep in a hold: the entire day seemed seeded in that dense, blue-white flare.

*

THE CURLING CLUB had a bar and lounge with finely-scratched Perspex windows overlooking the rink's four sheets of ice. Finding a space at the orange Formica ledge that ran below the windows, I pulled up a plastic stacking chair beside a man in his mid-thirties – burly, pale, clean-shaven, wearing glasses in black plastic frames bound at one hinge by a skin-coloured plaster, the man's skin paler than the plaster. Teams representing local businesses – the Port Authority, Hyska's Insurance, R&V Yamaha – gathered below us on the ice.

My neighbour leaned forward, elbows on the orange Formica ledge, chin resting on a plinth of laced fingers. He wore a cherry-coloured acrylic sweater and a blue wool hat, or toque. His features – narrow eyes, flat nose, thin lips – fought shy of making too big a splash in his wide face. We talked as curling got under way, the lounge filling up with supporters of one team or another, the air thickening with hubbub and the fug of crushed-in bodies and cigarettes. Sam had served as a mechanic with the Canadian Air Force; now he was a machinist at the port. He spoke with the diffidence of someone expecting a setback.

'This isn't the season for polar bears,' he said. 'You don't see too many tourists in Churchill unless they come for polar bears. I don't see why anyone would come to Churchill unless they wanted to see polar bears.'

'I'm looking for snow geese,' I said. 'I'm waiting for the snow geese to come up on their spring migration. I started off in Texas and came north more or less with the geese.'

'That's pretty far, eh? How long did it take you?'

'I've been travelling for a couple of months.'

'I guess snow geese'll be here soon. We've got Canadas here already, and snow geese come up soon after. So you like curling?'

'I've never seen it before.'

'Must look pretty strange. Rocks and brooms and such.'

'Are you a curler?'

'Me? No way. I don't even like it. There's not much to do round here. It's just good to get out of the house.'

A curling team, Sam explained, consists of four players, led by a skip. The game is played on a track or sheet of ice forty-six yards long from foot line to foot line. At each end of the sheet there's a tee surrounded by two circles, one red, one blue; the area inside the circles is the house, twelve feet in diameter. The player has to slide the curling rock towards the house, aiming for the tee, letting go before he reaches the hog line. Two of his team-mates can use brooms to polish the ice in front of the stone, clearing away grit, straw or ice crystals, the friction of the brushes generating heat, melting the ice, keeping it smooth and fast, or 'keen'.

Players were testing grey curling rocks, feeling for the weight of a stone by easing it forward and back, each stone shaped like an Edam cheese, buffed to a shine, glinting in the bright lights. The bar was crowded, stifling, rowdy with laughter and argument.

Sam and I went downstairs to the rinkside, pushing through a swing door, and I was instantly woken and pepped up by the chill off the ice, invigorated, as if I'd stepped out into the purged, clarion tone of high mountain air. We sat on a bench below a poster reminding players of the six 'essential' rules of curling, the top of Sam's toque right under the words *Each player must deliver when his turn comes*. Skips were squatting in the houses, shouting instructions; stones were rumbling down the sheets; players were pulling on flat-soled Bauer and Asham curling shoes, or handling Duke 8-Ender and Lowry's 100% Horse Hair brooms, or crouching, like runners in starting blocks, in the black rubber Marco hacks set in the ice behind each house. The chilled rink resounded with the tocks of stones caroming off other stones.

'Have you ever been on a snowmachine?' Sam asked.

'No.'

'It's pretty fun, I guess.'

He offered to take me out on to Hudson Bay; I accepted earnestly. He leaned forward, elbows on knees; he cupped his face in his hands, letting his chin rest on their heels. Now we gave all our attention to the curlers. Some used the strength in their arms to push the stone down the sheet; others tapped the power in their legs, kicking out from the hacks and sliding, stone in hand, eight, nine, ten yards up the ice, right up to the hog line. They slid until the friction on their gloves, shoes and jeans began to slow them down, and only then let go, eyes fixed on the stones as they sailed away towards the tees, plying the centreline or veering in sad, inexorable drifts to left or right. Sweepers kept pace with the rocks, scrubbing at the ice in front of their feet, skips shouting 'Up! Up! Up!' at them while players on neighbouring sheets scooted from house to house – one shoe slippered in a Teflon slider, the other with its tread exposed for traction – and waiting curlers knocked back bottles of Labatt Blue and Bacardi Breezer, rocks colliding musically with other rocks, in and around the houses.

*

FOR DAYS, GOOSELESS, without purpose, feeling increasingly lonely and foolish, I walked. People warned me not to walk along the coast, to watch out for polar bears: I never strayed far beyond the small woodframe houses. I walked to the elevator and Cape Merry, gazed out across the great white plain of Hudson Bay, then walked back down the railway line towards Winnipeg. I sat in Gypsy's Bakery drinking tea, writing in my notebook, or installed myself in the library, leafing through books, staring at the fibreglass polar bear that stood on the shelves, stepping from Fiction D–F to Fiction J–M. Winds howled in from the north; temperatures dropped to ten below. My hopes fell with them. I knew better than to expect geese in such conditions.

Solitary, adrift, craving familiar ritual, I ran through a white storm to the tiny Anglican church, its walls shipped to Churchill from England in 1890. North-easterly winds lashed in across the frozen bay throughout the service of Morning Prayer, catching the church broadside with cuffs and buffetings. There were tin panels inscribed with the Apostles' Creed and Ten Command-ments, a rectangular canvas painted with the glyphs of Cree script, and a wooden plaque engraved with a sentence from Ecclesiastes: 'Truly the light is sweet, and a pleasant thing it is for the eyes to behold the sun.' The church shook whenever the wind gusted. A congregation of four, huddled beneath a pendant electric heater, sang hymns with no accompaniment in this precarious fastness by the sea.

Ruth greeted me warmly in the anteroom. She was in her

fifties, a few isolated white strands in her neat black hair, chips of red stained glass hanging from her ears. She wore a chunky hand-knitted zip-up sweater with a flopping collar. Two Canada geese were knitted into the front of the sweater, facing each other across the zip. Ruth zipped it up, bringing the bills of the two geese together in a tender kiss, then nuzzled her chin in the neck of the jersey.

We sat down under black-and-white photographs of pioneer Anglicans. I said that I was from England; Ruth said that her parents had come to Canada from London in 1934. They had settled in Toronto, where her mother had found work as a chocolate dipper.

'She'd sit down just like this,' she said, 'with a marble board in front of her and all the centres ready to her left. The centres might be, I don't know, a nutty something or caramel or what have you. One of the servers would come by with a pail of fresh chocolate. There was one server for every five dippers. She'd ladle out some chocolate on the marble surface and my mother would swill it around a bit so it cooled down and she got it to just the right temperature. Then she'd pick up one of the centres and roll it once, twice, and whatever dripped off at the end would be that little peak you get on chocolates. Those liquid centres such as cherry would actually be hard before she dipped them, but the chocolate coating would be hot, which would melt the cherry centre. When she came home she smelled of chocolate. She always loved to eat chocolate, but later on she developed diabetes. She couldn't eat it any more. That was always something missing from her life, that she couldn't eat chocolate.'

The anteroom smelled of chocolate.

'What brings you to Churchill?' Ruth asked.

I told her about the snow geese.

'I see.' She didn't seem surprised. 'So you're just waiting around. Where are you staying?'

'In the bed and breakfast on Robie Street.'

'Well, look, I'm thinking out loud, but maybe here's an idea. My nephew's getting married in Tobermory, Ontario. That's on the Bruce Peninsula, between Lake Huron and Georgian Bay? I'm going to the wedding and I'm going to be away for a couple of weeks. I've been looking for someone to take care of my animals. You could stay in my house at Goose Creek on condition you looked after my dog and my cat. It's a good place to see birds. It would be a nice place to wait for geese in.'

My spirits lifted; I wanted to embrace her.

*

TWO DAYS LATER I drove Ruth to the airport in her pickup, a battered red Chevrolet Cheyenne. She was wearing a white sweatshirt printed with an English country cottage: a thatched roof; a garden brimming with roses and honeysuckle; a gate ajar. She said that the radio in the pickup was stuck on the country and western station, and I'd do best to switch it off and listen out for birds. Temperatures were above freezing now, the white storms all but forgotten. There were Canada geese in the thawing tundra ponds near the airport; the first herring gulls were up from the south. I left Ruth at the terminal and drove alone on the road to Goose Creek.

The road ran south, alongside the railway, metalled for a few miles, then narrowing into a rutted dirt track that entered a forest of weak-looking, stunted spruce trees.

The house was set in a clearing, a long, low cabin built of cedar trunks, planed flat and painted a pale matt red, with thermometers screwed into the doorframe, wind-cups spinning on the roof-angle: Ruth recorded their data in weather journals.

Her animals, my charges, were lounging outside the cabin, at the foot of the four stairs that led up to the door. Saila was three-quarters wolf and a quarter husky, with the colouring of a wolf (white legs and chest, black and grey on top) and the figure of a Shetland pony: fourteen years old, lame, deaf, almost blind, her dark eyes swirled through with milkiness. The cat, Missy, was slim, silver-grey, a paragon of stealth. I said hello to both, but they paid me no attention as I carried my bags up the stairs into the cabin.

Inside, like an old ship: very dark, with everything crowded in, shimmed cedar trunks varnished deep caramel brown, the shelves fitted with lips to prevent glasses and recipe books from falling should the house decide to pitch or keel. There was a heavy black Cummer wood-burning stove with a funnel running to the ceiling, a stack of spruce logs in the back porch. A black wrought-iron match-holder was screwed into the wood behind the stove, and all around it were paler scratches in the varnish where matches had been struck on the friction of the grain. Ruth had left me a note on a yellow Post-It saying, 'House may creak. It sits on tundra.'

Missy's playthings hung on cords from the ceiling beams: a yellow Sesame Street Big Bird and a fluffy purple mouse, dangling on strings of baubles and bells. Framed prints of wildfowl and hunting scenes hung on the walls: American avocets lifting from a pond, following the lines of their own upturned bills; hunters rowing out in a small boat to place decoys in a marsh, the sun rising behind them; four wood duck standing at the edge of a stream, with curling maple leaves strewn across the mud and water, the drakes painted like harlequins, sumptuous as geishas with their bright red eyes, white face stripes, iridescent manes, and imperial purple chests, 'wood duck' a plain name indeed for such a glitzy show.

Yes, the living-room had the character of a ship, with its shimmed logs, wood cabinets, lipped shelves, skillets and casseroles hanging on hooks, Aladdin mantle lamps with glass chimneys and fabric wicks, the brass ship's clock manufactured by

Hüger of West Germany, and the matching barometer, calli-
brated in millibars, conditions displayed in three languages:
Schön Fair Beau; Veränderlich Change Variable. There was a gas
cooker, a sink, and water in plastic drums by the spruce logs in
the porch, and behind the cabin was a patch of sodden, mossy
ground where squirrels and snowshoe hares ventured out from
the trees and an old door was laid as a ramp to the outdoor
privy. I unpacked my books and folders, piling them on the
table, thanking Ruth out loud, as if her house would hear me.

I was glad to have a base, somewhere to settle, establish
myself, gather my thoughts. Ruth had mentioned that she was a
crafter, a quilter and embroiderer, but I didn't understand what
this meant until I passed through into the corridor to look for
my bedroom. Fabric, everywhere: frills and ruchings, pelmets
and valances of pleated chintz, lace tablecloths, plush pull-
through rugs, flouncy knitted holders for plant pots and fruit
bowls, rag dolls slumped in corners, woolly ornamental swans
with drooping necks, tea cosies resembling igloos and tuxedoes,
bright-coloured furbelow in all directions – as if you'd stumbled
into a nest feathered with soft things and could throw yourself
anywhere you liked without fear, there would always be some
heap of feather bolsters or folded eiderdowns to cushion the fall.

The corridor, which Ruth used as her workspace, was almost
impassable for ironing boards; vises and pine planks for quilting
frames; heavy pinking shears; folders of stencils, templates and
patterns; a dressmaker's dummy with dials to swell or contract
the bust, waist and hips; boxes of spools, reels, thimbles, and red
party balloons to help grip needles as you pulled them through
the heaviest fabrics. Here were three sewing machines – a White,
a Singer, a Pfaff – and piles of gingham, flannel, calico and
velveteen, and white Dacron batting and polyester fibrefill for
the stuffing of quilts and cushions. And then gear for patchwork,
piecework and crazy quilts; felt angels and roses for appliqué;

hummocks of rag, lace, lint and old stockings; wicker trays of pine cones and blowzy artificial flowers; and reels and rolls of cotton, nylon cord, jute twine, raffia straw, twill tape, bias tape and worsted-weight acrylic wools. I was smiling even before my eyes came to rest on the jars of gaudy bobbles, pompoms, shiny doodads, fake feathers, sequins and glitters of silver and gold, and the upbeat embroidered proverbs that grinned out at me from scrolling frames: *Happiness is a Warm Iron; Quilting Forever! Housework – Whenever!*

I reached my bedroom, the last room in the cabin. Ruth had put clean sheets on the bed and covered it with a sky-blue quilt, and on top of the quilt she'd placed a pair of homemade pillows shaped like angels' wings – two wings of shining white velveteen, each feather a purse of fibrefill sewn shut with gold thread, arranged correctly, heel to heel, spread wide against the sky-blue, as if inviting me to lie back and feel their fit between my shoulderblades. There were billowy white sheers in the window, and generous pleated curtains of floral chintz. The walls were hung with tapestry pictures of mallard and Canada geese, and with silk scarves tied in cascading and butterfly bows, and on either side of the bed were luxuriant pull-though rugs in which my feet sank almost to the ankles. I put down my bags, rejoicing. Ruth's cabin was more than a home. It was a world.

*

I WROTE AT THE TABLE in the living-room. Saila slept on the red carpet in front of the Cummer stove. Her chest heaved. She snored and twitched. Her legs no longer hinged at the knees: they were as stiff as crutches. Each step forward beat the odds, bucking a trend. She tottered. She moved one foot and waited before following its lead, as if to verify that the limb could still support her weight. She staggered into the bedroom every morning between four and five o'clock, nosing me awake.

I'd open my eyes to her big wolf's face, dawn light caught in the sheers. I'd get up to let her out, and we'd walk down the corridor together, one step at a time, blind Saila listing from side to side, slewing into stacks of boxes. One morning she knocked over a box marked 'Christmas Decorations' and stood confused in a spill of rosettes, pompoms and papertwist angels.

I spooned tins of beef flavour TriV dog food into Saila's dish at the foot of the steps and measured out biscuits in an old margarine bowl. Missy ate tins of Friskies. Her dish sat on the sideboard, next to the sink, on top of a back issue of *Country Woman* magazine, open at the *Readers Are Wondering* page. Victoria Soukup of Iowa had written: 'My hobby is collecting turtle memorabilia. Does anyone have adages involving turtles to share?' Pam Simakis of West Virginia wished to learn to make patio lights from plastic flowerpots and toy biplanes from empty soda cans. Charlotte Lovegrove of Indiana asked, 'Any ideas where I might obtain a player piano roll of the song *Back Home Again in Indiana*?'

Each day seemed warmer than the last. The wind had turned: now it blew from the south and south-west, a wind for migrants. Herring gulls arrived in increasing numbers. The ice started to break up on the Churchill River, softened by warm water flowing from the south. I walked with Saila along the tracks of Goose Creek as the thaw set in. Rivulets and pools appeared in the open ground between the tracks and the spruce trees. Moss clumps and sedge tussocks glistened with moisture. The air was rife with clicks: the furious Morse of wood frogs and boreal chorus frogs emerging from hibernation. Saila yawed from one side of the track to the other. She stopped for minutes at a time while her strength welled up again and the frogs persisted with their Geiger din.

One day, as usual, Saila woke me up at dawn. I let her out and stood for a moment on the steps of the cabin. Dreams were spilling over. Frogs were clicking in the tundra pools. And from the direction of the river, faint but unmistakable, I heard the calls of snow geese: twenty-five or thirty birds, blues and whites in about equal proportions, yapping like terriers, flying in a waving oblique line over the spruce trees towards the cabin in the half-light. Saila was blind and deaf but she noticed the flock, lifting her head from the waterbowl as the geese passed low above us, their wings thrumming like voltage. The dog went back to her drink. I went back to bed.

*

SAM, WEARING HIS blue wool toque and brown Bollé sunglasses in place of his broken black spectacles, was standing on the ice at Cape Merry beside two Polaris skidoos, styled yellow and black, like wasps. The ice in the mouth of the Churchill River, from the Cape to Munck's Haven and Fort Prince of Wales, was rough and creamy, ocean chop modelled in plaster-of-paris; close to the shore it was buckled and heaved up in messy ructions by the action of the tides.

'This ice you got holding tight to the shore is shorefast ice,' Sam said in his soft voice. 'About a quarter mile offshore you've got a lead – a band of open water between the shorefast ice and the floe. Shorefast ice doesn't move, except up and down on the tide. The floe ice moves in and out on the wind. We're going to head out to the lead. That'll be what we call the floe edge.'

The sky was a clear, deep, voluminous blue. The skidoos were easy to operate: brake and throttle, no gears. The handles were heated. To cut the engine you hit a red button marked with a zigzag lightning bolt. We yanked the starter pulls.

I followed Sam out of the Churchill River to the ice of Hudson Bay. The going was awkward close to shore, but once we'd passed the Cape the ice was smooth and flat, with shallow meltpools and patches of slush that sprayed from the runners of the skidoos in sparkling fantails. We accelerated out into the bay, outracing the fumes of the engines. I tried to keep to Sam's tracks, skirting faultlines, guiding the runners round potholes in the crust. Sam raised his right hand, the signal for a stop.

'There's a smaller lead just ahead,' he shouted, competing with the engines. 'Two or three metres of open water. Keep your speed up and we'll skip right across it. Just go straight at it. Don't let your speed drop.'

We circled back to get a run at the lead. I watched Sam's skidoo splash across. Then I thumbed the throttle hard against the handle; the skidoo bore down on the open water; the hull bounced once on Hudson Bay before the curved tips of the runners, upturned like avocets' bills, found solid ice again. We sped on out to the belt of open water separating the fast ice from the floe. The strength of the silence stunned me when we cut the engines. The silence was all the more intense for the context of the engines, and it took time for the ear to make out subtler sounds, the delicate undermusic of drips and tricklings, like the chinking of fine broken glass, as sinuous meltwater rills ran through the crust and the corners of iceblocks yielded, drop by drop, to the thaw.

The floe had driven again and again into the fast ice: the edges of both had buckled, heaving up reefs, boulders, plates and menhirs of packed snow. We stood at the edge of the lead. Sam's brown Bollé sunglasses had side panes to cut out the glare on the flanks. The ice forms on our side were glistening, with blues and blue-greens ghosting in the whites, sunlight glancing off wet edges and corners. The tide was going out, the ebb streaming underneath us, the open water turbulent with eddying and upsurge. Across the lead, ice ramparts on the edge of the floe were cut with the precise shadows of cornices, overhangs, abutments and scarps.

'I was out here this winter,' Sam said, speaking softly again.

'Forty below. Strong winds blowing from the north. My god, you could see the floe driving into the fast ice, bulldozing right in with the wind. Parts of it reared up and buckled, but there were other parts where the plate of the floe was actually lifting and crawling up over the fast ice. It moved real slow, with a noise like tyres spinning in a snowbank. When the floe lifted over the fast ice, brine flooded up from under. Seawater swilled up and froze right in front of me. My god, I watched it turn white. The water rushed up and suddenly it was ice.'

A ringed seal surfaced in the lead, eyed us for a moment, then swam away, ducking and cresting, sleek and pliant, like liquorice. Six Canada geese flew over us towards Fort Prince of Wales, their black necks, white chinstraps and pale rumps showing clearly against the empty sky, and then a pintail drake passed in a fast, skeeting glide along the lead, its tailfeathers long and pointed, a white scimitar line curving up its neck and head. A flock of snow geese, ten white-phase and five blue-phase birds, followed the lead to the north in a loose chevron, with a lone sandhill crane behind them, legs trailing, neck stretched without a kink, distinct from the pleated neck of a heron. I saw the crane's bill open wide a second before I heard its magnificent shriek. My senses were reeling with ice gleam: the heady, implacable grandeur of a frozen sea.

*

MORE AND MORE BIRDS were arriving from the south. Ducks gathered in the tundra pools around Goose Creek. They muttered while the frogs clicked: the background uproar had texture now. Ruth had collected waste grain from the elevator and left heaps of it on the mossy ground behind the cabin. I kept my bird books and binoculars by the sink and tried to identify the passerines: American tree sparrows with rust-coloured heads and a dark spot on their breasts; white-crowned sparrows with zebra-striped scalps; dark-eyed juncos remembered from David's bird-table at Riding Mountain.

One morning a truck pulled up outside the cabin. I heard the door slam, feet on the four stairs, two knocks on the door. An elderly man was standing in the porch, holding up a purple string net bag containing three bulbs of elephant garlic, produce of Chile, purchased at Spice World in Orlando, Florida.

'I'm looking for Ruth,' he said.

George came from Nevada, Iowa. He had smart Polaroid glasses, a grey moustache, and a grey fedora, a shade darker than the moustache. He was devoted to birds. Since retiring from the US Navy, he'd made the trip to Churchill once a year to photograph migrants. I was explaining that Ruth was attending her nephew's wedding when George spotted something in the yard and reached for the binoculars on the sideboard.

'I'll be darned!' he exclaimed. 'If it isn't a cowbird! A goddamn cowbird! Can you believe that! Knock me down with a feather! A cowbird!'

A brown-headed cowbird – a kind of blackbird with a brownish hood and a bill like a finch's. Like the European cuckoo, the cowbird is a brood parasite, laying its eggs in the nests of other birds and leaving these foster parents to rear the young cowbirds as their own. When we looked up the cowbird in the field guide, I understood George's excitement. The guide showed the winter and summer ranges of each species by means of shaded bands on maps of North America no bigger than postage stamps, and though the cowbird's breeding range broached southern Manitoba and extended quite far north in Saskatchewan, it never got close to Hudson Bay. This cowbird outside the cabin at Goose Creek was either a pioneer or it was lost: a vagrant, like Gallico's snow goose.

'So Ruth's in Tobermory, huh? And you're here to look after the animals?'

I told George about the snow geese and my idea of following them from one home to another. He reacted just as he had reacted to the cowbird.

'I'll be darned!' he exclaimed.

We went out to watch birds, driving slowly in the Cheyenne along the tracks of Goose Creek, binoculars round our necks, maintaining the vigilance of sentries. We wound down the windows, admitting the tremendous frogsong. The tundra between the tracks and spruce trees was sodden, mossy, bristling with tussocks, glittering with meltwater pools. There were ducks everywhere. George and I raced each other to call out the names.

'Pintail!'

'Shoveler!'

'Scaup!'

But George was the expert. He identified a flock of five bufflehead from their swift, low flight; I remembered their white head-dresses from South Dakota. He pointed to a circling northern harrier and knew that up close you'd see it had the round face of an owl. He spotted a new gull, fresh from the south: Bonaparte's gulls, much smaller than herring gulls, with thin straight black bills instead of the heavy down-curved yellow bills of herring gulls, and wings angled at the elbow like a tern's. George knew that bonies were named after Charles Bonaparte, nephew of Napoleon; that they wintered along the Atlantic, Pacific and Gulf coasts of America, and bred in a wide band from Alaska across the Yukon to James Bay – these gulls may have flown north-west from New England, or straight up the Central and Mississippi Flyways with the snow geese and cranes, or on the long haul from Orlando, Florida, like the elephant garlic. We saw an American kestrel perched on the nib of a spruce, flocks of snow geese in familiar undulating skeins, the black necks of Canada geese poking above sedge tussocks.

We left the truck and walked, stopping now and again to

raise binoculars and scan the tundra. I spotted a wader, and George recognized it immediately as a lesser yellowlegs, grey-speckled, slender and elegant, with a long thin bill like a pipette, its light body raised high on straw-yellow legs, up to the shins in meltwater. The calls of snow geese some distance away made a faint tinkling like wind chimes or marinas where the halliards are pinging on metal masts, and the honking calls of Canada geese seemed to sound on both in-breath and out-breath, like harmonicas or the rubber black-bubble klaxons of classic cars.

'Wow,' whispered George. His binoculars were trained on the tundra. 'Look at that.'

I raised my binoculars. I tried to steady the lens, as a sniper would. In the open ground were sedges, heaths, moss clumps, rivulets, small pools, thickets of shrubby willow, the browns of single-malt whiskies and old sellotape.

'Bitterns,' whispered George.

There they were: two American bitterns, *Botaurus lentiginosus*, cryptic, secretive birds, a kind of egret, standing motionless, practically invisible in their brown, black-stippled plumage, their necks stretched, their bills, which had the scabbard shape of herons' bills, thrust skywards. The bitterns had noticed us and were pretending to be reeds, trusting that the genius of their posture and colouring would conceal them from view.

'Oh my,' whispered George. 'Aren't they beautiful?'

The birds didn't flinch. They might have been whittled from wood, or chanced upon in the thick of a bole. Birders call them thunder-pumpers on account of their booming territorial calls, but this pair were committed to a strategy and wouldn't risk a whimper.

'Wish I had my camera,' whispered George.

Later, alone again in the cabin, with Saila sleeping by the Cummer stove, her body twitching, as if in the grip of last throes, I referred to field guides and found the bittern's pumping call rendered in the peculiar phonetics of birdbooks. *Oong-KA-chunk!* one offered. *Oonk-a-lunk!* another transcribed. I tried to imagine the sound of a bittern. I concentrated on the phonetic renderings. But how did you convert the signs to actual song?

*

RUTH'S CABIN, Sam's gentle friendliness, George's passion for birds: these were gifts. Without them, I might easily have lost heart. In the last bitter days of April, trudging back and forth through Churchill, I'd passed the travel agent on Kelsey Boulevard again and again, thinking how easy it would be to go in, buy a ticket, fly home. In Gypsy's Bakery I'd made lists of things I had to do in England, as if I were already there. I'd thought of friends I would see, haunts I would revisit, foods I would eat, pieces of music I would listen to. And I'd nursed resentments against the place I was in, with its white storms and frontier harshness, the elevator bearing down like a force of oppression.

Most of all it was the cabin that lifted me, that restored my energies and refreshed my enthusiasm for the journey with snow geese. I had somewhere of my own, a secure base. Forms were distributed in fixed, reliable patterns. My feet learned the width of rooms, my hands the location of handles and switches. I settled into routines, unchanging from one day to the next – getting up at dawn to follow Saila down the corridor; pouring water from the yellow plastic drums into the sink for washing-up; spooning Missy's Friskies into her bowl on the sideboard, Mrs Amos King of Pennsylvania seeking Strawberry Corelle dishes and other strawberry-themed items. I dusted Ruth's framed proverbs and family photographs. I confided in Saila. I cooked comforting, familiar dishes. I slept with the curtains open, just in case I woke in the middle of the night with an aurora right outside the window, diaphanous and spectral.

One night I dreamed of the ironstone house. But it was in

the wrong place. It wasn't in the middle of England, beside a wood, with farmland sloping upwards to the south and west, a spire poking the sky to the north. The house was on the edge of Hudson Bay, with caribou and polar bears wandering in the garden, the windows filled with dazzling white: pure floe.

How could I not think of home, when so many birds were homing overhead? When I'd set out, going back had been too far off to contemplate, beyond the horizon, not visible for the curve of the sphere. But now return seemed imminent, within reach, as if I'd gained a coign of vantage and saw my home range in the distance, a day's walk. Only Foxe Land lay between me and that known world. I was restless again, restless for the known. I felt the draw of the familiar, as if I'd entered a field of gravity. I wanted deep attachments instead of fleeting encounters. I wanted the things I saw and heard to accord with the things I remembered. I wanted roots. And all these were waiting beyond the Great Plains of the Koukdjuak.

Homesickness and nostalgia no longer refer to the same condition. Both describe states of return-suffering, but while the homesick individual longs to go back to a particular place, to return in space, the nostalgist longs to go back to the past, to return in time. Both can be a medium of escape. You might dream of going back to some idealized place or time because you are frightened or unhappy in the here or now, as I had been frightened and unhappy in hospital. Then, it was as if illness were itself a foreign country, where nothing was recognized or understood; as if being ill were a kind of expatriation, a forced removal from conditions to which you had become habituated and attached. Going home was at least a kind of going back. Home was a reprieve from the unpredictable.

I wanted to guard against such fantasies of escape. I couldn't rush back to the old ironstone house whenever circumstances outside it became inhospitable. Nowhere was my sense of belonging as complete or unambiguous as it was in my child-hood home, but if I saw that sense of belonging as something exclusive to the ironstone house, then I would never really leave, never grow up, never look for my place in the world. Somehow I had to turn my nostalgia inside-out, so that my love for the house, for the sense of belonging I experienced there, instilled not a constant desire to go back but a desire to find that sense of belonging, that security and happiness, in some other place, with some other person, or in some other mode of being. The yearning had to be forward-looking. You had to be homesick for somewhere you had not yet seen, nostalgic for things that had not yet happened.

One afternoon I drove the Cheyenne along the coast to see the wreck sitting just offshore near Bird Cove. The shorefast ice had yet to break up, but further out, beyond the lead, the ice was already in pieces, with shards of blue sea showing between the floe hunks. I took the metalled road towards Cape Churchill, turning off down a track, pulling up close to the edge of Hudson Bay. I didn't walk far from the Cheyenne, scared of polar bears hidden in the tundra hollows. Snow geese flew high overhead in loose, flexing chevrons. In front of me, gripped by ice, was a ship, rusted all over, decaying, its sad-looking cranes and davits and broken funnel distinct against the blue and white. It had been driven aground, caught in a windstorm in September 1961 while carrying nickel ore from Rankin Inlet to Montreal. I stood by the Cheyenne, panning across the wreck with my binoculars, flocks bleating high overhead. But all I could think about was the ship's name.

The *Ithaca*.

*

THE NUMBERS OF snow geese flying over the cabin at Goose Creek began to drop in the second half of May. Many had already settled at the nearby La Pérouse Bay breeding colony; many were flying further north, up the coast of Hudson Bay, then swinging east to Southampton Island and Baffin Island. After walking past it so many times, thinking about going home, I pushed open the door of the travel agent on Kelsey Boulevard and booked myself on a flight to Baffin. The next day I picked Ruth up from the airport and drove her back to the cabin. I'd already said goodbye to Saila. Ruth said she wanted to pay me for looking after her house and animals, and I said I wanted to pay her for letting me.

Sam and I met for the last time in Gypsy's Bakery. His smile wasn't conventional; it was a kindly scowl. Sometimes he took off his blue wool toque and scratched his head, revealing thin, wispy hair, but never for long. He was wearing an old leather bomber jacket, a T-shirt just tight enough to hold a pack of cigarettes against his upper arm, and a gold watch strapped so the face was on the inside of his wrist: to tell the time, he had to turn his palm upwards, rolling the forearm, the face just where a nurse would feel for a pulse, his body's timekeeping adjacent to the watch's.

He'd suggested we drive out to Cape Merry. The sky had been clear all day, with waterfowl and passerines rafting in on clement, moderate winds from the south. Sam's pickup was parked outside Gypsy's – no ordinary pickup, but a Ford F100 from 1956, painted bright scarlet, with natty chrome plaques on both doors (*Custom Cab* in a sloping script) and genuine running boards you could hitch a ride on if the cab were full. The seat was a banquette, sprung like a trampoline, covered with a brushed cotton sheet bearing faded motifs of pink flowers. Ruts and frost heaves set passengers bouncing. On the rear fender was a sticker that read, *If you don't like the way I drive – Stay off the sidewalk!* and Sam had glued a Canadian Air Force shield to the dashboard, with its motto *Per Ardua ad Astra*. Through striving, to the stars.

'I've just got to get something from the shed,' he said. 'Then we'll drive out to the Cape.'

Sam's shed was a small free-standing garage next to his bungalow. Inside stood a Harley-Davidson and a gyrocopter. The gyrocopter had no wheels or rotors: it was an elaborate cage, standing on the points of its tubing, as if on tiptoe, with a black seat resembling a child's car safety seat lodged in its centre. An old guitar case, wrapped with packing tape, leaned against one wall, close to a pair of Rossignol cross-country skis with boots clamped in the bindings. Over the workbench a hand-written sign said, *Clean up better than you found it, OK?* and above that was a long plywood shelf with a lip where Sam had written: *Fragile! Gyrocopter Rotor Blades!* He rummaged around for a while, then hoisted a green kitbag over his shoulder.

The elevator blocked out the low sun like a censor's mark. Sam threw the kitbag into the back of the Ford. The steel bodywork was left bare in the cab and smelled of wet stone. Sam had bought the truck when he was with the Air Force in Camp Gordon, Ontario. He'd bought the Harley-Davidson in Winnipeg. Churchill had about thirty miles of paved roads: Sam cruised them as if they were the grandest highways.

'You hit the end of the road and turn around and go straight back again, and on and on like that until you're sick of it or out of gas,' he said. 'When I left the Air Force, I rode the bike from Thompson, Manitoba, on the Dempster Highway north to Inuvik, and that's as far north as you can drive. I had a tent. At night I'd pull off the road wherever I found myself. I'd push as far as I could into the bush and camp there, and I guess I found a river to wash in every three days or something.'

We passed the elevator, leaving the asphalt for the uneven rubble track that led out to Cape Merry. Sam parked the Ford at the end of the track. We clambered over boulders covered in pale green, grey and rust-red lichens, making for the tip of the promontory, stopping at a wide, flat rock right by the river-mouth, the ice nearest us forced up in sooty rifts where it hinged on the rise and fall of tides. Sam pointed across the river to the remains of Fort Prince of Wales, the old Hudson Bay Company stronghold, the sun on the edge of the bay to the north-west.

I asked Sam about the gyrocopter.

'That's my current project,' he said. 'I'm building it myself. I'm using ideas from about six other gyrocopters, just taking the best ideas from those and putting them in my own design. I bought one basic set of blueprints plus I've had other design plans from pictures. You can order the equipment in from the States; it's pretty easy, actually.'

'Where are you going to fly?'

'Oh, I don't know. Just around.'

There had been other projects. Sam had made a power kite from a design in a book. He'd used an old hang-glider sail, and doubled the proportions recommended in the plan.

'Yeah, I made an extra-large version of it. Then I brought it out here and went out on the ice on my cross-country skis. Wind wasn't bad. Pretty soon I had the kite launched. It pulled me along at a lick across the fast ice, the smooth part before it's all smashed up close to the lead. I was holding on as tight as I could to the handles, and the wind was just screaming into the kite. It pulls you along real fast, faster than you want to go. I had a friend on a snowmachine who came and picked me up after each run. You see, I wasn't able to use the kite back into the wind, to tack on it like you would in a sailboat.'

For several years Sam had kept a sailing boat in Vancouver, moored in Burrard Inlet across from Stanley Park. He sold the boat after almost sinking it off Bowen Island. A violent squall caught him sailing alone.

'The boat was sideways on the water for about three minutes,' he said. 'I didn't know how to get the sails down. Everything was jamming up on me. I went into brainlock. I couldn't understand what was happening, it was all so sudden. I was trying to pull in a roller furling and keep hold of the tiller. The squall was driving the boat sideways, and I just got lucky it blew over so fast. I really thought I was going to die. Yeah, I thought I was really close, and I just felt sad. I didn't feel much other stuff I'd expected to. Just sad, mostly.'

We sat down on the rock, looking out into the bay. I imagined Sam speeding past on the ice, clutching kite strings, his toque a blue spot whizzing by. His projects were variations on a theme: schemes of escape.

'I like to get lost on purpose,' he said. 'I like to get deliberately lost. I got lost on purpose in China, Russia, Mongolia, the Philippines. I got lost in Beijing and wound up in an industrial section in the middle of nowhere. I got deliberately lost in Manila. I was in Germany with the military, doing NBCW training, nuclear biological chemical warfare training, and got deliberately lost in Berlin and Munich, and other places I don't even remember the names of.'

A pair of Canada geese flew low from left to right in front of us, keeping to the layer of cool, dense air above the ice. There was no wind. Snow buntings flitted from rock to rock. Three larger white birds appeared on the rocks below, stepping about on the lichens: willow ptarmigan in their snow-white winter plumage – a kind of grouse, with a chunky body, short legs, round black eyes and a small hooked bill. Ptarmigan are year-round tundra residents: their plumage chameleons with the season, white moulting into checkered chestnut browns, the birds remaining camouflaged even as the snow disappears. They have feathers right down their legs, like trousers, and develop feathers on their feet in winter (their scientific name, *Lagopus*, comes from the Greek for 'hare-footed') both to keep them warm and to function as snowshoes, reducing the distance each foot sinks in snow by as much as fifty per cent. These three didn't seem bothered by two men sitting on a rock.

It was about 200 yards across from the cape to the fort. There were shallow meltwater pools in the ice. Sunlight flickered on the water like candles under panes of glass. We could feel the sunlight on our faces. Sam gestured towards the rivermouth, stretching out his left arm the way people often do to draw a cuff back from a watch.

'I swam across there,' he said.

'I believe you. I'm not sure I'd believe anyone else.'

'It was just a challenge. This was 1988. I wasn't long out of the military. People were pretty negative about the idea. Nobody here much wants to see anybody who stands out for any reason. I trained for a year. I knew I needed to swim as fast as possible across the river because of the cold, so I trained myself in the front crawl. Nobody ever taught me to swim. I taught myself when I was a kid. I started long-distance swimming when I was stationed at Baden Baden in West Germany. They had a pool there, but more often I'd swim in the quarry behind the base. There was a deep quarry pool with cliffs around it. It was somewhere you could be on your own, I guess.

'I started training as soon as I'd decided to swim the rivermouth. I swam in the pool all winter, at least an hour a day. I swam miles in the pool. When the ice broke up on the lake in June I swam in that to get used to the temperatures. I'd go out there on my own and grease myself up with Tenderflake lard, which is a white beef fat. I bought blocks of it from Northern Stores. I slathered my hair with the lard, and put a bathing cap over that, and wore swimming goggles. I don't think the lard helped too much, but I'd read that people who swam the English Channel used that.

'I trained all of 1988 and the first half of 1989. I did the swim on August the first 1989. The river's usually open from

July through September. I tried to pick a day when the water was as warm as I thought it was ever going to get, but it was still really close to freezing. I paid a couple of guys boat money to come with me in a boat. Nobody came to watch. I'd told a few people but they weren't really interested. I started right here, on the Cape, swimming across towards the Fort. The thing was to start off right at the bottom of low tide. In the afternoon, maybe one o'clock. I greased up with lard and pulled on the bathing cap and swimming goggles. You have to start swimming as soon as you hit the water. You can't waste any time. As soon as you're in the water your energy level's plummeting. It's in free fall. It doesn't come back if you stop to rest. The water's so cold it's sucking the life out of you, and the only thing for it is to just keep going.

'The thing is, on the swim across, as soon as you put your head underwater, you can hear the whales. There are lots of beluga whales in the river at that time. I swear you can hear them perfectly. These are white whales. People call them sea canaries. They make a whining sound, it's like a psychedelic whining sound, unlike anything you ever heard before. One of the guys on the boat said that while I was swimming he saw a beluga whale swim right underneath me.

'It took me half an hour to swim across to the Fort. It was much harder getting back. I got swept up in these tidal currents which took me five hundred feet out into Hudson Bay. I knew I had to keep swimming. I swallowed some water and cramped up almost immediately. My goggles fogged over. I couldn't see where I was going. I followed the sound of the boat. It was an aluminum boat with an outboard, and I just kept swimming for the drone. Then for some reason the guys slowed up, and I almost swam right into the propeller. I swore at them because I was sinking. I was doing a zigzag course back in from Hudson Bay and I was pretty close to passing out. I think I was hallucinating a little. The whales got to sounding like mermaids and I thought maybe I should head down and join them.

'Eventually I got back to the Cape. When I got out I didn't know where I was, I was so hypothermic. I felt like I was looking way down at my feet. My feet were a long way down below me. I was shaking violently and almost convulsing. The guys from the boat sat me on a chair and covered me with blankets. They gave me hot tea, and I started to warm up a little, and I guess I was happy. Before I did it I was either going to do it or I was going to die, that was my thinking. I trusted myself that I could do it. When I warmed up a little, I had this feeling of gratitude. I wanted to thank the spirit of the river, for letting me live.'

We sat quietly for a while. Sam took off his toque and ruffled his hair. Then, abruptly, he stood up.

'Do you want to shoot some potatoes?' he asked.

We walked over the rocks to the pickup, and Sam retrieved the green kitbag. Then we went back to our flat rock with its prospect of the river, the bay, and the sun, orange-gold, low on the horizon.

The kitbag contained a large plastic contraption resembling an exhaust pipe: a homemade gun, a howitzer fashioned from the kind of black PVC tubing used for guttering and downpipes. The barrel was striped with yellow tape, and the trigger, an electric barbecue lighter, was fastened to a red wooden handle with black tape. Wires led from the trigger to a chamber in the butt of the gun.

From the green kitbag Sam produced a plastic Northern Stores bag full of potatoes. He pushed a potato down the muzzle of the mortar, ramming it home with a dowel, working fluently, like a musketeer. There was a black line on the rod, with arrows pointing towards it. Next to the arrows, in black felt-tip, Sam had written, *Potato up to here*. He unscrewed a cap in the butt of the gun and wafted fresh air into the chamber.

'Would you pass me that can of Lysol?' he asked.

I found the disinfectant in the kitbag. It contained denatured ethanol and had a 'Crisp Linen' scent.

'Lysol! The choice of champions!' mocked Sam.

He sprayed Lysol into the chamber for a few seconds, then screwed back the cap. When you pulled the trigger, the barbecue lighter would spark the gas in the chamber, causing it to combust.

'I'll shoot the first one, OK?' Sam said.

I stood aside. Sam aimed the mortar out over Hudson Bay. He pulled the trigger. With a roomy *whoof* like wind buffeting a small woodframe church, the explosion launched the potato on a long, high trajectory. We tracked the missile – a dot, a speck, then nothing, falling invisibly, somewhere in the ice, close to the sun. We were laughing.

Sam took another potato from the Northern Stores bag and pushed it down the barrel with the dowel. He unscrewed the cap in the butt. Smoke poured out of the chamber. He wafted in fresh air.

'You've got to have enough oxygen in there,' he said.

He sprayed in more Lysol, screwed back the cap, and handed me the mortar. I held it low. I pointed the barrel at the sky over Hudson Bay and squeezed the trigger. I felt the explosion at my hip, heard the potato whoosh up the tubing. Another arc: dot, speck, nothing.

We took it in turns, shooting potatoes till the bag was empty.

8 : FOXE LAND

WE TEND TOWARDS HOME. Migrant birds don't travel for the sake of it. They move between winter and breeding grounds because the Earth's axis is not perpendicular to the plane of its orbit round the sun. They migrate in response to the tilt, to the seasons and seasonally variable food supplies that exist on account of the tilt. In any species, an individual that remains within a familiar environment has more chance of finding food and water, more chance of avoiding predators and exposure, than an individual that strays into unknown territory. Homesickness may simply have evolved as a way of telling an ape to go home.

When I climbed down off the plane in Iqaluit, Baffin Island, my first impulse was to look for birds: herring gulls, skittering snow buntings, ravens with slick vinyl lustre in their feathers. The sky was stuffed with cloud, like the down filling in a duvet, snow tuning the light to its own white pitch. Painted prefabricated houses made rudimentary occasions of colour and line in this whiteness. Dented, rusting pickup trucks, and four-wheeler ATVs blazoned with the names Big Bear, Timberwolf and Kodiak cruised the reddish-brown dirt roads. Figures trudged through rubble and dirty snow along the roads' shoulders, huddled in heavy green or navy knee-length Arctic parkas with short reflector stripes on their breast pockets and fur-trimmed hoods. Caribou grazed on patches of bare ground. Teams of huskies, roped together like mountaineers, slouched on the ice of Frobisher Bay.

It was 23 May, the day my father expected swifts to arrive back at the house. On Baffin, my green insulated Below Zero gumboots, Capilene thermals, fleece overgarments and puffy down-filled jacket were little proof against the cold. I lugged my bags from the airport to the Discovery Hotel, regarding the Arctic parkas with envy and covetousness. The night was dark for an hour or two, and I slept fitfully, confused by this extended photoperiod. By morning the cloud had vanished. The sky was deep blue, flawless; I needed to visor my sunglasses with a flat hand, as if saluting, to protect my eyes from the fierce Arctic albedo.

During the night a gas explosion had blown out the back of an apartment block in Happy Valley. A crowd gathered, families gazing up at interiors rudely exposed to open air: cookers, televisions, velveteen sofas, dishevelled beds, the edges of floors and walls ragged with torn carpet and mangled steel, a thick black smoke column rising into the blue. Firefighters trained hoses on the smouldering apartments, visors pushed up on their yellow helmets, gas tanks inverted on their backs, valves not at the neck but at the base of the spine, as if to distinguish them from scuba divers. One crew member stood watching, resting: a woman with matted hanks of black hair on her forehead, her cheeks smirched with charcoal. She was smoking a cigarette, staring at the burning house, taking long, deep drags and breathing out grey smoke, thinner and lighter than the smoke churning upwards with constant, liquid volume from the burnt-out apartments.

Nearby, the Cathedral of St Jude was shaped like a snow-house, with windows and a cross at the zenith of its white dome. Inside, behind the altar table, hung purple drapes covered with bright-coloured appliqué decorations: snow geese, beluga whales, walruses, dogteams, igloos. And in front of the drapes hung a marvellous cross: two narwhal tusks, presented by the community at Pond Inlet, fixed to an upright and crossbeam of varnished oak, cusps and furrows spiralling round each tapering ivory shaft from hilt to tip. The lectern was a sled stood up on its feet, with a bible open on the sloping topslat, between the curve of the runners, and a microphone fixed in the prongs of a fish spear.

I remembered another church, its spire the first landmark you'd glimpse above the trees, marking the site like a golfer's flag, a skip's broom held upright in a curling house, as the mast stayed by taut steel hawsers marked Matthew's house in the hills outside Austin – a jutting stone pulpit, white marble cartouches, pendant electric heaters glowing orange over deep-toned oak pews, hollows worked into the floor's stone slabs by centuries of shoes. The lectern was a brass eagle; a bible lay open like another bird on the backs of its wings (an eagle, children were told, to wing the words to heaven); a snow-haired widow played the organ as if driving a freight truck, deft with pedals and dashboard instruments, checking a rearview mirror for the clergyman's signals.

*

THE SMALL JET entered thick cloud over Meta Incognita Peninsula and emerged again over the southern edge of Foxe Land. Far below, the ice of Hudson Strait was breaking up, with stretches of open water making black slicks in the white surface, and polygonal plates of ice drifting across larger, lake-sized clearings in the floe. I kept my face pressed to the window, hoping to see flocks of snow geese flying into Foxe Land on the south-west winds, wondering if the white birds with black-tipped wings would even be visible in these monochrome vistas. I felt the blood-hum of quickened expectation, the same excitement I'd experienced driving the blue Chevrolet Cavalier from Houston to Eagle Lake, and again as I waited for my first sight of snow geese at dusk near Jack's house on the prairie. I was about to see snow geese in Foxe Land, in their breeding grounds – perhaps even the very same snow geese I had seen in their

Texas winter grounds, in the long-limbed company of American white pelicans, great blue herons and sandhill cranes, three months and 3,000 miles ago.

Jeff was waiting for me at Cape Dorset's airstrip: mid-thirties, with a stocky wrestler's build, a thick Russian-looking moustache, and hard brown eyes like two hazelnuts behind little round wire-rimmed glasses. The rarefied scholarly fineness of these spectacles sat incongruously with his rugged outdoorsman's figure and belligerent geniality. He was standing by a Big Bear ATV, smoking a cigarette with impatient, muscular purpose, wearing quilted waterproof dungarees, insulated black Sorel glacier boots and a green Arctic parka with a fur-trimmed hood. His chin and cheeks were covered with chestnut bristles one third the length of his moustache bristles, and this stubble looked like a protective accessory, a sensible adaptation to extreme wind chill factors. He worked for the Department of Renewable Resources and had offered to help me find snow geese.

'Hey!' he said, as if throwing a punch.

'Jeff.'

'Damn right!'

He heaved my two bags on to the rack in front of the handlebars of the Big Bear, sucked the last life from his cigarette, threw the stub like a dart at the dirt track outside the airstrip building, then straddled the four-wheeler, bouncing on its suspension.

'Get on,' he said. 'Hold tight. Don't let go.'

I climbed on and we ripped down the dirt road, the afternoon light heavy with the whiteness of cloud and snow. The buildings of Cape Dorset were distributed over hills below us: prefab huts and houses, panels of colour raised several feet off the ground on steel piles sunk through permafrost to the bedrock. Inuit children ambled up and down the mud-brown track, calling out as the ATV sped past, and sometimes Jeff accelerated towards them, cackling with villainous theatrical gusto, my grip tightening round his waist as he leaned forward over the handlebars, streamlining the Big Bear, the children soon recognizing their nemesis and stepping aside, ceding the run of the road.

We stopped at a two-storey prefab wooden house, stilted like a bayou house, with matt clay red walls, a shallow-gradient roof of Vic West corrugated steel, and icicles hanging from its eaves in the orderly, staggered lengths of tubular bells: the icicles would sing different notes if you struck them with a hammer. A large, thick-haired husky was tied to the short flight of wooden steps leading up to the front door.

'Shooter! Have you missed me? Poor baby! Poor Shooter!' Jeff said in a mewling voice, squatting down, ragging the dog's ears, taking its lavish pink licks on his bristled chin and cheeks. I followed him up the stairs into the house.

'Home sweet home,' Jeff said, unzipping his parka.

The spare room was furnished with a narrow bed and a simple wooden table. There were no decorations on the walls.

'I really appreciate this,' I said.

'Don't sweat it, buddy,' Jeff replied. 'Put your bags down. Let's get out there and take a look.'

We rode back up the hill on the Big Bear, following the dirt track up above the airstrip to the tank farm: large white drums holding diesel, gasoline and aircraft fuel. Jeff switched off the engine. There was no perceptible dimming of light to tell you it was evening. We left the four-wheeler and walked out across the snow.

'Keep your head up,' Jeff said. 'Don't want to get caught by a bear now.'

My body was struggling to cope with the cold, my feet almost numb in the green Below Zero gumboots, my hands lifeless in thick gloves. We walked laboriously, the snow crust apparently firm under our feet, promising a solid platform, only to collapse as soon as it had gained our confidence, plunging us knee-deep or thigh-deep in soft snow, making each step a cameo of optimism ending in disappointment. Jeff walked as though he had a bone to pick with the snow, his short sturdy legs ramming down with the force of pistons – as if he'd knocked the ground to the ground and now intended to stamp the life from it. We reached a vantage point and stopped.

Whiteness. Below us, prefab housing units around the harbour, the water covered with snow ice. A few roped-together dogteams lay on the white surface, doodles on a blank page. Beyond the harbour you could see the open water of Tellik Inlet running out into Hudson Strait. The inlet remained ice-free all year, a stretch of water known as a polynya. The surface was motionless, without whitecap or wrinkle, a deep-saturated, viscous grey, a sea asleep. Beyond it snow-covered hills merged with white sky, land distinguished from sky by the black flecks and stipplings of rocks.

'Igneous granite,' Jeff said. 'Been here since the beginning of time.'

We walked a few yards to the left and stepped up on to an outcrop of black rock. The black on the granite's surface looked like a charring, but in fact it was a lichen, a living tar, the black overlaid with patches of other pale green and yellow lichens, growing out from their own centres like tie-dye patterns. I stretched out my left arm, drawing my coat's cuff back from my watch. It was ten o'clock, evening, but the light still held to the idea of day, with no sign that night was imminent or ever expected. The light was hyperreal, flushed with precious metal, platinum-tinged.

Jeff lit a cigarette, cupping the flame in hands as burly as boxing-gloves, then pointed across Tellik Inlet to the rolling white hills.

'Sometimes I'm out there,' he said, expelling words along

with cigarette smoke, 'I'm out on the land, and it's like the void. It's like a sentence or two *before* Genesis.'

'I'm going to have to find some thicker gloves,' I said. 'Maybe a thicker coat.'

'You think this is cold?' he asked. 'Eight or nine below? This isn't cold. This is *Hawaii!*'

He laughed, throwing his head back, a manic, gleeful, booming laugh, as loud as he could make it, as if hoping to leave traces on the muffling white. But Jeff's laughter hardly broached the silence. He sucked on the cigarette. We looked out over Tellik Inlet. There was no wind. A few gulls flew low across the open grey water. The huskies lying roped together made simple patterns on the harbour ice: points joined by lines. The silence was something you could hear, as though it were itself a sound: a steady, white drone. The light was suffused with the character of blades and foil. Across the water, chains of hills merged into white sky with no visible seam.

Jeff and I noticed the geese simultaneously – a faint horizontal line, flying northwards: a skein of fifteen or twenty birds, black wing-tips flickering above the hills, the line undulating as if a lazy energy were rolling through it from tip to tail; blue-phase and white-phase snow geese, borne across Hudson Strait on the south-west winds of the past few days, back in Foxe Land after a winter sojourn close to the Gulf of Mexico. I felt a sudden lightness in my chest. We'd got here on the same day.

Jeff threw down his cigarette and raised both arms in the air, celebrating.

'Whoa!' he yelled. 'Geese are coming! Look at those babies! Oh man! We're going to have us some geese!'

'Good timing!' I said.

'Damn right!' Jeff was ecstatic, restless on our rough granite podium. 'Oh man. I've got a good mind to go straight over the other side of the island right now. Waste no time, pick up a gun, start shooting. We're going to bake a snow goose for you real quick.'

The flock trailed northwards, over the mainland.

'I love you!' Jeff hollered at the birds. 'My babies!'

*

IN THE LAST WEEK of May, waiting for the chance to go out into Foxe Land, I walked the dirt tracks of Cape Dorset as I'd walked the length and breadth of Churchill – getting my bearings, working up a mental map, configuring the settlement according to certain prominent landmarks: the Northern Stores, the Dorset Co-operative, the cairn-topped hill that marked the harbour's entrance. The community, which numbered about 1,000 people, mostly Inuit, was spread across three valleys on the north side of Dorset Island, just off the southern edge of the Foxe Peninsula. The prefab wooden housing units had small windows to minimize heat loss, and this made them look like people squinting.

Skidoos and long sleds called qamutiiks were pulled up by the front steps of the raised box-like houses. Stretched sealskins dried on the porches; the stiff tailflippers of flensed seals poked from rusting oildrums. Inuit children careened down the winding dirt roads on weather-beaten ATVs as the ancient municipal sewage truck hauled itself from house to house. Someone had parked a Yamaha Enticer snowmobile outside the Lillian Prankhurst Memorial Full Gospel Church, and on the long qamutiik hitched to it stood a glittery purple Ludwig drum kit: bass drum, snare, high-hat, cymbals – a cymbal-clash of colour.

I walked back and forth, following the lanes and paths, experiencing the same views again and again, until they became familiar prospects. I got to know the layout of the place, how forms were organized within it, but I couldn't shake off my sense of disorientation and estrangement. My eyes wouldn't relax into the hyperreal, silver-lined light, the whiteness bundled like sheets and towels into a space too small to contain it all. My circadian rhythms were confused by each day's failure to darken. People spoke Inuktitut, a language I couldn't begin to understand. I walked around town bewildered, off-balance, nervous, detached. Now and again a husky sat up on the harbour ice, pointed its nose skywards and crooned, each open-throated *harrooo* quickly gathered in by the low cloud.

*

I FOUND JEFF sitting on the linoleum kitchen floor, legs akimbo, plucking feathers from two blue-phase snow geese with brisk tugs.

'So,' he said, not looking up. 'You ready to go out on the land?'

'More than ready,' I said.

'You better go soon.' Little muffled puffs of white goose down appeared wherever Jeff yanked flight and contour feathers from their follicles.

'I want to.'

'There's no way to travel once the thaw sets in. It's way too wet out on the tundra. The snowmachines just get stuck, or sink, more likely.'

'I can't wait to get out there. I've been travelling with the geese for quite a while now. I'm about ready to head home.'

'Don't sweat the small stuff, buddy. You know? I guess the thing for you to do is meet the guys at the HTA. The secretary speaks some English.'

'That's what we talked about on the phone. The Hunters and Trappers Association.'

'Right. You can meet the elders. Don't let them fool you,

though. They understand more English than you think. Just
don't sweat it, OK?'

*

THE DIRECTORS OF the Aiviq Hunters and Trappers Association met once a week in a small shed behind the Dorset
Co-operative building. The secretary was a wiry man, all liga-
ment and sinew, with heavy steel-framed glasses and long black
hair hanging down below his shoulders. His hair had the rough
fineness of the hair in violin bows, and he wore a black anorak
with the NFL logo on its sleeves and the word RAIDERS on its
back, just visible beneath a curtain of dry, fanned hair.

'Welcome to Cape Dorset,' he said quietly, surreptitiously,
confiding a secret.

Three of the elders were already seated, drinking coffee.
Posters depicting the ecology of seals, walruses, polar bears and
the salmonid fish called Arctic char were pinned to the walls
alongside large sheets of paper on which ideas for future HTA
projects had been scrawled in black marker pen. One said: *Goose
Down. Collection? Cost? Needs study*. Three more elders came
into the room and took their places at the tables. Five of the
six were men, late-fifties to mid-sixties, with weathered, high-
cheekboned Asiatic faces, dressed in jeans, worsted lumberjack
shirts, and baseball caps embroidered with the name of the new
Canadian territory, Nunavut, or the logo and slogan of the Polaris
Mine: *Five Years Accident Free!* The sixth elder, sitting on
my right, was a robust woman in a blouse on which twining
green foliage, apples, pears, grapes and ripe, plump plums were
displayed in Arcadian abundance. Her long black hair, silkier
than the secretary's, was drawn back from her face and held
at the neck in an elegant silver barrette; her cheeks gleamed
under the weak electric light as if coated with a slipware glaze.

The secretary distributed copies of the meeting's agenda. The first item was 'Prayer'. The elders closed their eyes and looked down while the woman sitting beside me said a prayer in Inuktitut. When she had finished, the secretary addressed me in his low voice.

'Please tell everyone where you come from,' he said, 'and why you're here in Cape Dorset.'

The elders looked at me expectantly.

'I'm from England,' I said. 'I'm interested in snow geese. I've followed snow geese on their spring migration from Texas. I've come to Cape Dorset because I want to see the geese in Foxe Land, in their summer breeding range.'

The secretary made notes on a pad, then translated for the elders. The word 'Texas' stood out like a flaw in his Inuktitut. The man sitting to my left doodled on his agenda: spirals, arabesques, a neat head-and-shoulders sketch of a man in a fedora or panama. The wooden tables, like school tables, were scored and notched with graffiti, the wood splintered where ballpoint pens had gouged against the grain. The secretary finished speaking. There was a pause. The woman addressed the gathering in Inuktitut, smiling, cheeks gleaming. The elders laughed. The secretary turned to me and translated.

'She says that you need to be careful about who takes you out on to the land. Some of the young people, they think they know the land, but really they don't. You don't want to get in trouble out there.'

The doodling elder looked up from his agenda and addressed the directors.

'He says that travel is getting hard,' the secretary translated. 'Spring's coming. Sometimes it can be too wet. You can't cross the rivers any more. The snow isn't hard enough for the snow-machines.'

Then all the directors began talking at once in Inuktitut. Sometimes they laughed. Finally, the secretary turned to me again.

'OK,' he said. 'So. We're going to ask around and see if we can find someone who's going out on the land, who'd be willing to take you. You're staying at Jeff's. We'll contact you there and let you know.'

'Thank you,' I said.

'So we'll contact you at Jeff's,' the secretary repeated.

I took the hint, stood up, nodded at the elders and left the meeting.

*

MAY BECAME JUNE. The top of the Earth was tilting towards the sun. Nights were betrayed as nights by a slight slackening in the light, a sullen humour in the sky. I struggled to keep track of each day's passing, without the grand events of sunrise and sunset as points of reference. I learned to distinguish white land from white sky by the flecks of black granite, but a raven holding steady against the wind could quickly confuse the issue.

The sky cleared; sunlight knifed off the snow and ice. Late in the evening I'd walk out across the harbour, passing dogteams asleep on their ropes, keeping to skidoo tracks where the snow was fretted and compacted underfoot, then turn and look back at the community, its districts nestled in hills: to the left, Itjurittuq, or Roman Catholic Valley; to the right, Kuugalaaq, the Valley; in the middle, Kingnait, or Town. Weak-coloured, impermanent structures, without foundations, raised on thin stilts like wading birds.

One evening I walked to the hill beyond Itjurittuq, trudging through deep snow to the summit cairn, looking down on the strandline where the harbour's white ice fronted the open water of Tellik Inlet. The grey sea surface lay heavy and tight, with a

few small bergs drifting across it. Dogs barked. Children shouted.
These sounds rolled round the bowl of the hills like balls in a
roulette wheel. Snow buntings flitted from rock to rock below
me, their black-and-white plumage entirely of a piece with the
prospects beyond them.

I sat by the cairn. The gleam off the ice and water was blue-
tinged, this blueness like the nuance carried by a remark, an
inflection or emphasis to the way the light was speaking. Gulls
preened on hunks of broken floe in Tellik Inlet, heads down,
fiddling in their armpits. A seal surfaced. Skidoos tore across the
harbour ice, plying the diagonal from Kuugalaaq to Kingnait. A
flock of king eider flew from the south, skimming low over the
water, breasting their own reflections, easily distinguished from
geese by their rapid wingbeats, the flock's quick kaleidoscopings.
King eider winter along the Atlantic coast from Labrador to New
Jersey; many breed on Baffin Island. More new arrivals, returning
to their natal grounds, their journeys almost over.

Five Inuit boys clambered through the snow to the cairn,
four of them ten or eleven years old, in parkas, baseball caps,
baggy jeans and black Sorel glacier boots, and the fifth much
younger, tagging along, wearing a blue wool toque.

'I keep my cigarettes in there,' said one of the boys, pointing
to the cairn.

'What do you smoke?' I asked.

'Players Light. You?'

'Nothing.'

'You speak Inuktitut?'

They told me a walrus was *aiviq*; a bear, *nanuk*; a snow
goose, *kanguq*.

The boys began foraging round the base of the cairn, trowel-
ling at the snow with half-cupped hands, gathering stones. They
chose flat stones, as if for ducks and drakes, though we were too
high above water for skimming, then launched one missile after

another, each limber loose-armed pitch accompanied by a grunt of effort, the stones hanging above the white mainland hills before dropping to the blue-grey surface of the inlet. I was sitting just a few feet from the boys but I watched them as if from a distance, or as if the scene were being projected on to a screen in front of me, my senses registering the silver-toned alien light, the water below, the glistening bergs, the hills rolling away in infinite white vacancy, but my mind elsewhere. The stones rose spinning and faded like clay pigeons fired from traps, and I imagined gunshots, targets exploding high above the water, designs for fireworks.

*

'WE'VE FOUND SOMEONE willing to take you out on the land,' said the secretary. We were standing outside the Dorset Co-operative building. The secretary, in the black NFL anorak and heavy steel-framed spectacles, pushed his long hair back behind his ears, tracing small sickle curves with his fingers. 'The woman at the meeting? Paula? She's going hunting. She's going with her son, Natsiq. She says you can go with them. OK?'

*

JEFF LENT ME a can of CounterAttack bear-repellent spray and an old green Arctic parka: a quilted, down-filled, knee-length coat with a horseshoe of light grey coyote fur round its hood, a map of dark oil stains on its back, and several badges of grey masking tape repairing tears in the fabric. Jeff drew my attention to the label on the parka. The coat was called a Snow Goose. I tried it on, engulfed, surprised by its heaviness on my shoulders: it was like wearing a bungalow.

In the West Baffin Co-op I bought six cans of naphtha fuel for Paula's Coleman stove, four bottles of Yamalube oil for the snowmachines, and a chit for twenty gallons of gasoline, and carried these supplies in two loads to Paula's house in Itjurittuq. I was wearing my green Below Zero gumboots, three pairs of socks, Capilene thermals, fleece leggings, waterproof Gore-Tex trousers, two fleece tops, inner and outer gloves, a blue wool toque and the Snow Goose parka. In my bag I'd packed binoculars, a notebook, extra clothes, and my grandmother's copy of *The Snow Goose*, with Eleanor's auburn hawk's feather like a bookmark between its hard blue covers.

Paula and Natsiq were checking the Enticer and Polaris skidoos. Paula, long black hair pulled back from her glazed cheeks and cinched at the nape in an elastic band, wore an unzipped green Snow Goose parka, navy-blue waterproof dungarees over a bright red polo-neck, and black Sorel glacier boots. Her son, Natsiq, was about thirty, slight, dark-complexioned, with high, jutting cheekbones, a wispy beard, and dense, furtive eyes. He was dressed in a black fake-leather bomber jacket, black waterproof trousers, black Sorel boots, and a baseball cap embroidered with the word *Aksarnerk* – Inuktitut for 'northern lights'. He chain-smoked du Mauriers, lighting one cigarette on the ember of its predecessor. Paula nodded at me and said something to Natsiq in Inuktitut. Two qamutiiks were loaded with orange-red fuel cans, bedding rolls, food boxes, ammunition, radio equipment and guns, all lashed down under ropes, bungees and blue tarps. Natsiq took my bag and lodged it securely in the qamutiik hitched to the Polaris.

'Ready?' he said.

He yanked the Enticer's starter pull and heaved on a navy Snow Goose as the engine turned over in a low chug. He slung a Maverick pump-action shotgun across his back – the gun rusted all over like a salvaged musket – and straddled the skidoo, revving the engine. Paula started the Polaris, zipped up her parka and slung a rifle across her back – a Ruger fitted with veteran telescopic sights, its steel barrel bent out of whack. She indicated that I should get on the Enticer behind Natsiq. I held tight to his waist. The Snow Goose gave him a bear's girth. We moved off down the dirt road to the harbour, skidoo tracks rumbling on grit, the runners of the long slender sleds scraping on chips of stone. The sky was clear; the albedo fierce. Behind my sunglasses I squinted like the houses receding behind us.

The skidoos and qamutiiks took to the harbour snow as if returned to their native element. Natsiq opened the throttle; we tore past huskies lounging in tethered teams; we raced breakneck across the clean white plane to the hills of Mallik Island. I could feel the Maverick's stock and breach pressing into my chest, the ash from Natsiq's du Maurier peppering my face in the icy apparent wind as we sped away from Dorset, cresting the first slope and sailing as if on our own momentum down into Foxe Land, the whiteness around us flecked with black granite protrusions, the parka's heavy fur-trimmed hood bouncing on my shoulderblades, the sky not a finite canopy but a blue opening-out into ever larger spaces, and my heart pounding, roused by the exhilaration of our speed, the light's brilliance, the nearness of snow geese in their natal homes. We cut eastwards over the mainland and then hit the sea again, accelerating across the hard roof of Hudson Strait, skidoos and qamutiiks rocking on fissures in the shorefast ice, the snow blown in gleaming dune-smooth ridges called sastrugi and smaller windrows like the fins of white Cadillacs caught in the freeze.

For two or three hours we travelled along the south coast of Foxe Land, short-cutting across peninsulas and headlands, with flocks of snow geese and Canada geese passing overhead, from right to left, coming in off Hudson Strait on the south winds: contingents of twenty or thirty birds, arranged in their limited alphabet of V, U, J and W formations, white-phase and blue-phase family groups intact in each skein, black wing-tips beating with inked distinction on the clear, vivid sky. Inland, in the valleys, flocks were grubbing in patches of open tundra, crowding the first available districts of sedge and running water, and the engine-roar of the two snowmachines sent geese up in flurry after flurry, as if the land's own surface were breaking loose.

Sometimes Natsiq and Paula stopped the skidoos, and Natsiq unslung the rusty Maverick, crept up as close as he dared to a feeding flock, then shot at geese. He shot two blue-phase snows and a Canada, securing the dead birds under bungees on the qamutiiks.

Once, out on the ice of Hudson Strait, we stopped to examine footprints in the snow: deep, round, five-toed impressions in the crust, the prints of a polar bear, nine or ten feet from nose to tail, travelling alone, coming in off the sea. Once, racing eastwards, a lone tundra swan flew low over our heads, heading in the opposite direction, its broad-winged cruci-form shadow slipping across the open ice. And once, we stopped a mile offshore for no obvious reason, the two machines tiny on the brilliant ice plane. Silence – the steady, white drone of it – poured into the space vacated by the engines. The flat white surface stretched away into mist like the edge of the world. Natsiq pointed towards the mist. Looking along the line of his arm, I found the fat blackness of a ringed seal, slouched beside its breathing hole, a dollop of life. Natsiq crept up on the seal, hunched over, keeping low. He got down on one knee and raised the Maverick to his shoulder. I trained my binoculars on the seal. I held my breath. A gun-crack; an explosion in the snow like a puff of goose down. But the seal had slid down the breathing hole, unharmed.

We continued along the southern shore. Cairns and other stone figures stood on every promontory and eminence, making dark vertical nicks on the white ground or deep blue sky. These were inuksuit: *inuk* means 'human being'; *inuksuit* are stones that convey information as if a man or woman were standing in the landscape, most serving as aids to navigation, guides for hunters. My eyes were drawn to them, signs of human passage in the wilderness, holding the country in place like rivets or tacks. But the skidoos kept streaking eastwards, across the sea, the sibilance of front skis and sled runners just audible in the guttural engine roar. Above us, a blue abyss, with no flaw or blemish to give it scale, as if outer space began at the surface of the Earth. The snow seemed to touch directly on the void.

At Andrew Gordon Bay we turned north, heading inland, leaving Hudson Strait behind us. Snow and Canada geese rose off the valley floors, gaining height rapidly, veering from the two skidoos. But I'd stopped registering them as geese, as birds. My life offered no precedent for these surroundings. The new rushed at me, too much of it and too fast to be absorbed and processed. The motion, the albedo, the strangeness of ice plane and tundra, the numbness in my hands and feet, the constant engine roar – it left me dazed, dreamy, dumb, and for a moment it was as if we were travelling across the middle of a page, with whiteness and black markings all around us, and geese lifting off the snow like letters coming unstuck.

*

WE CAME TO a broad frozen lake, Lake Angmaluk, a small snow-roofed fishing cabin on its far shore. Natsiq and Paula parked the snowmachines outside the cabin and cut the engines.

Startling, the way silence romped in. Natsiq unhitched the bungees and we set to work, unpacking the qamutiiks. The cabin had plywood pressboard walls, a narrow sleeping platform, and a shelf that didn't quite run true, with a jar of instant hot chocolate, a can of cooked ham and a packet of White Lily Marven's Pilot Biscuits left on it. Two plastic travel mugs bearing the logo of the Pauktuutit Inuit Women's Association hung on nails below the shelf. Natsiq and Paula exchanged a few words in Inuktitut and told me by means of simple gestures where I should put things: foam mats, bedding rolls, the Coleman stove, the cans of naphtha fuel.

Paula took a blue-phase snow goose from the qamutiik, severed its left wing with a hunting knife, and swept the cabin's pressboard floor using the wing as a brush. Leaving the wing-brush on the sloping shelf, she sat outside, spread-legged on the snow, the rest of the goose between her knees. Muttering, she rent the carcass with a hunting knife, reached in with her right hand, and pulled out the bowels and craw. Purple-red offal steamed where it rested on the snow. Paula melted snow in a dented cauldron and boiled the skinned, limbless goose, adding a sachet of Lipton onion soup to season the broth, soon slicked and beaded with yellow subcutaneous fat previously stored by the goose as fuel for the long flight north.

We ate, sitting on the snow, leaning back against the shack. I didn't want to eat snow goose. I'd often imagined myself in Foxe Land, at the end of my journey, seeing snow geese return to the country in which they'd hatched. It had never occurred to me that I might have to eat a snow goose. I was attached to these birds. I couldn't help thinking of them as my companions. But I didn't want to set myself apart from Paula and Natsiq. So I kept quiet while we ate goose. A few down feathers floated in the broth. I took small mouthfuls, chewing solemnly. The meat was rich: you could taste the miles in it. The morsels sat in my stomach like pebbles. We ate in silence, snug in our parkas, snow geese inside and out.

Natsiq stood up and looked around, taking stock. He had a twitch, a way of puckering his lips and pushing them across his face towards his right ear, stretching the skin of his left cheek and waggling the point of his featherweight beard. He looked down at me. He was wearing the *Aksarnerk* baseball cap and pitch-black sunglasses with curving black leather side-pieces, the cabin and dazzling snow reflected in both round lenses. He stroked his beard with a gloved hand. Sometimes his twitch carried the beard's point outside the remit of the stroke.

'We go,' he said to me. 'Shoot geese. Hunting? Bang bang?' He raised his hands, miming, jerking upwards with each shot.

I didn't want to shoot snow geese, but I followed Natsiq down the shallow valley, away from Lake Angmaluk. He was toting the Maverick over his shoulder; my binoculars bounced against my chest just as they had on the walk alongside Sand Lake to Houghton Dam, when Rollin had talked about bald eagles and his daring flight beneath the Golden Gate Bridge. The snow crust was firm underfoot and I liked walking: it got my blood moving; it restored sensation to my toes. We came across the splayed hoofmarks of caribou, tufts of grey-brown caribou fur (the reindeer were shedding their winter coats), and whole tracts of snow patterned like a convict's uniform with the three-pointed footprints of geese. Flocks of snow geese and Canada geese flew overhead, yapping and honking.

We hid in the lee of a craggy granite slab, close to an area of exposed wet tundra. Natsiq removed his gloves, lit a du Maurier, loaded red shells into the Maverick and hunkered down, keeping out of sight. I followed suit, curling up on the snow, face against stone, the granite pink with iron in places, run through with crystalline quartz seams. I could see black, green and rust-red crustose lichens, and brittle, foliose lichens in paler colours, and clumps of green moss among chips of grit and quartz, with new buds forcing through, tiny pods of moisture and pigment like the vesicles in limes, and I breathed in the rich loamy vegetable smell that came off them, surprised, delighted, nostalgic for plants.

Natsiq took long drags on his du Maurier. Ash flakes caught in his beard. Together we listened hard for geese, tracking a

flock's approach by the crescendo of yaps or honks. Natsiq
waited until the flock was right on top of us, then got up on
one knee, raised the Maverick and started shooting, the butt
thudding back into his shoulder, red shell casings leaping over
me in short, smoking arcs, while overhead, against the deep
blue, a flock of Canadas or lesser snows veered away, calling
frantically, flocks breaking up, bevies of geese peeling off in
the fleurs-de-lys of aerial displays, circling, gaining height, and
gathering again, out of range.

Not all of them escaped. A white-phase snow goose dropped
to the snow like a cloth bag filled with coins. A winged blue
sheered off from its V, flapping in vain, down-strokes getting
no purchase on the air. I was shaking. I'd been shaking since
the shooting began. I left the slab's shelter to pick up the dead
birds, holding them by the shins, heads dragging on the snow,
a bright scarlet daub on the white bird's breast. I thought of
the young girl, Frith, carrying the snow goose to the hunchback
in his lighthouse hermitage. I tried not to look at the dead
birds, not to think about the dead weight in my hands. But
I kept shaking as we tramped back up the valley to the cabin,
carrying six geese between us, flocks still passing overhead on
the south winds. Whenever Natsiq looked up, tilting his head
back, puffing on a du Maurier, I saw skeins and chevrons
slipping across the black lenses of his sunglasses.

The light's tension had slackened: it was night. The Coleman stove was going inside the cabin. We washed down White
Lily Marven's Pilot Biscuits with mugs of instant hot chocolate.
Paula and Natsiq said little. It was a hunting trip, there were
things to be done. Paula rolled up my Snow Goose parka and
stuffed it inside a pillowcase, making a pillow. We got into our
sleeping-bags and lay back on the small platform, side by side,
Paula in the middle. The plywood pressboard next to my head
bore a message, written in careful ballpoint capitals.

ZEKE SLEPT HEAR
DREAMED OF CINDY
FACE CHAPPED — SNOWING

I lay on my side, considering this haiku. I wondered about Zeke and Cindy. Had it been snowing while Zeke sheltered in the cabin beside Lake Angmaluk, or in the dream in which Cindy had appeared? Whose face was chapped, Zeke's or Cindy's? Paula and Natsiq fell asleep beside me. My mind wouldn't shut down. It was thronging with geese.

*

NEXT MORNING we loaded the qamutiiks and left Lake Angmaluk for the Kimmik Hills. Spring seemed to have happened overnight. Patches of sodden, marshy tundra appeared along the valley floors, bristling with sedge tussocks, veined with trickling meltwater rills. Sunlight sparkled off rills and meltpools; the snow, softening, glistened with fresh, watery sheen. The going was hard for the skidoos. Paula and Natsiq stood up in the saddles to guide the skis round rocks or bogs, but as soon as we hit furlongs of flat snow they opened the throttles and we sped deeper into Foxe Land, Canada geese flying low alongside us like outriders. Sometimes Natsiq pulled up the skidoo to shoot at geese, and sometimes Paula tried her luck with the Ruger, creeping on stalwart bandy legs towards willow ptarmigan. The barrel's curve hooked each shot several feet from the target, and the grouse, in dapper feather spats and trousers, stood their ground unperturbed as bullets ricocheted off the granite around them, Paula's curses flying close behind, delivered with more accuracy than the bullets.

We travelled for just over four hours. Everywhere I looked, there were geese: flurries of Canadas and snows lifting from open tundra; flocks coasting overhead in familiar, straggling lines on the south winds. My hands and feet went numb. I held tight to Natsiq's waist, still dazed, watching geese, thinking of the 3,000 miles behind me. I wondered if any of these flocks had wintered close to Eagle Lake, or been counted by Michael at Sand Lake, or gleaned for leftover grain on the Portage Plains while the Viking, jeans secured with braces and belt, styled his steel-grey hair with nimble flicks of a pink comb. Sometimes I turned my head, looking back over the hood's coyote fur trim at Paula – her antique snow goggles, her face's beaming eggshell glaze, the Ruger's barrel poking at an angle from her right shoulder. Each qamutiik's load of dead geese grew steadily, bird by bird.

Natsiq stopped the Enticer on a ridge. We looked down on the white oval of a frozen lake, a broad valley stippled with granite protrusions and open tundra, and soon we were striking camp on a granite plateau beside the lake, raising a dun canvas tent, knotting guy ropes to a ring of small boulders, rolling out yellow foam sleeping-mats over damp black and green lichens. We sat round the Coleman stove. Paula melted snow in the dented cauldron; we ate cereal bars, Pilot biscuits and pots of instant noodles, all washed down by strong, sweet coffee. Mother and son exchanged occasional phrases, Natsiq stroking his beard, smoking one du Maurier after another. We rested for a while, then heaved on our Snow Goose parkas and left the warmth of the tent to walk around on the bare rock, the white lake immediately below us on one side, the undulating flecked white landscape of Foxe Land receding into mist on the other.

Slipping the can of CounterAttack bear-repellent spray into my pocket, binoculars hanging round my neck, I walked alone up the ridge behind the tent to a summit marked by a small inuksuk. I kept checking for bears, looking left and right as if preparing to cross a road. I could hear snow geese, their calls second nature to me now. My boots sank deep in the snow. I reached the summit. The inuksuk was a large piece of granite, crusted with black lichen, standing on end like a rough-hewn column, with two flat pieces of granite set on top of it. The pedestal might have been righted by the traction of a retreating glacier, the two topstones placed by melting ice.

I didn't touch it. I didn't stand too close to it. Some inuksuit marked the sites of deaths, murders, betrayals and acts of bravery; others had healing powers, contained spirits, expressed joy, happiness, evil or terror. There were inuksuit to indicate places of power, places where one should never sleep, places where human beings had been eaten, places of confusion and disorientation, and fearful places where travellers got lost. Some made whistling sounds when wind blew through them; some were healing arches where shamans effected cures. There were inuksuit to which one should show respect, inuksuit which couldp confer blessings and ensure safe passage. Certain inuksuit marked entrances to spiritual realms from which one returned unburdened.

Wind riffled through the fabric of my parka, the fur-trimmed hood luffing on my shoulders in stronger gusts. Far below, our tent looked like a jousting pavilion, wanting knights, pennants, escutcheons. There were hills and swales in all directions, drawing away on the curve of the sphere, and clouds massing in the south, above the sea. Wind blew into my ear as if into a shell, making a sea sound. A pair of tundra swans flew north, at eye level, their long lancing necks tipped with black bills. An arctic fox trotted across the frozen lake, a tinge, the ivory-yellow of old piano keys, its coat changing from white to a blend of tundra browns as summer approached. Flocks of snow geese pressed northwards. Through binoculars I tracked a pair of white-phase birds as they glided down to a stretch of open tundra beside a stream. They'd start to build nests as more and more tundra was exposed, choosing sites on top of slight undulations, where the ground was comparatively dry and firm.

I wouldn't see those nests: in two or three days the tundra would be too wet for the snowmachines. I wouldn't reach the Great Plains of the Koukdjuak, just over the horizon to the north. I was resigned to this. I didn't mind. I was in Foxe Land, with snow geese. Exhilarated, light-headed, I stood on the summit, next to the inuksuk, breathing deeply. Apart from the wind, all I could hear was geese: the faint halliard tinkle of distant flocks, the sharp yaps of nearby birds, the low electric thrum of beating wings.

Natsiq was sitting outside the tent, smoking a du Maurier.

'Tomorrow,' he said. 'Kingnait.'

I nodded.

'Home!' Smiling, tic twitching his wispy beard.

*

PAULA BOILED another snow goose. She left the Coleman stove burning while we ate. Muzzy, head swimming, drunk on the fumes, I saw myself hurtling backwards, southwards, everything in fast-motion, a plane flying tail-first from Iqaluit to Churchill, the Muskeg Express rattling back to Winnipeg, the Greyhound reversing all the way down Interstate 35, a film rewinding. I remembered driving the blue Chevy Cavalier from Houston to Eagle Lake, the mesquite trees and shambling longhorn cattle, the galvanized farm sheds and rice bins. I thought of snow geese flying from those Gulf Coast prairies to Baffin Island, flying according to inherited programmes, modified in adult birds by the experience of previous journeys, determining their direction by reference to the sun, the stars and the Earth's magnetic field, pushing north at the leading edge of spring.

In August, prompted by Zugunruhe, they would fly south again, just as swifts would be flying south across the Mediterranean to their African winter grounds. In a few days I might be watching swifts. I was high on naphtha fumes. I was brimming happily with the fact of being here, with geese, in Foxe Land, on the brink of return. I was ready to go back. But I didn't want to go back to the conditions of childhood. I didn't want to feel safe inside the old ironstone house. Not all returns are retreats, and if I wanted to go home, it wasn't a dream of escape, it was because love can't exist without the pain of separation, and so much of what I loved was there.

Resting my head on the rolled-up Snow Goose parka, I opened *The Snow Goose*. Eleanor's hawk's feather fell on to my chest. 'The bird was a young one,' I read, 'no more than a year old. She was born in a northern land far, far across the seas, a land belonging to England. Flying to the south to escape the snow and ice and bitter cold, a great storm had seized her and whirled and buffeted her about. It was a truly terrible storm, stronger than her great wings, stronger than anything. For days and nights it held her in its grip and there was nothing she could do but fly before it. When finally it had blown itself out and her sure instincts took her south again, she was over a different land and surrounded by strange birds that she had never seen before.' I remembered Mr Faulkner reading to us in the high-windowed classroom, women gathering at practice tees to loosen up their swings. I closed the book. There was a pale grey light inside the tent. Paula and Natsiq were asleep beside me. But I couldn't sleep. Even now, I could hear geese.

*

WHEN MY SPIRITS had been low, alone in the white motel room, adrift in Churchill, confused in Cape Dorset, I'd looked ahead to the moment of return, willing it closer. Now I wondered when my going back began. Was I already going home when I walked down to the tent from the inuksuk? Or not until the next morning, when we packed the qamutiiks and straddled the skidoos, my gloved hands linking at Natsiq's chest, the engines turning over? We moved away from the campsite, a ring of stones recording our tent's circumference on the granite plateau, and soon we were heading south, scudding down the lowland tundra swales. I was going back. My mind seized on the word, repeating it – *back, back, back* – until it lost all meaning. Geese were rafting overhead on the wind. My hands and feet went numb again. There were heaps of dead birds in the long sleds.

We reached the sea that evening: low, massed cloud; pewter-tinted light; the threat of storms. We kept to the shorefast ice of Hudson Strait, hurtling westwards into thick fog. I still felt the pull, more powerful now, of an intimate gravity, as if that force and not the snowmobile were conveying me homeward, not a hard fall but an easy sailing back towards the centre. And when I glimpsed figures standing on the ice, vague in the fog, I assumed I'd dreamed them up, delirious, mirage-ready. But there *were* people: fifteen or twenty Inuit, five families coming home from a fishing expedition, in Snow Goose parkas and Sorel glacier boots, chatting, smoking cigarettes, stamping their feet to keep the blood moving, fixing the ski on a snowmachine or topping up fuel or securing ropes on qamutiiks, with children running between the sleds, playing tag in the gales that howled in off the Arctic Ocean. My hood was up, my parka zipped as far as it would go, and I looked out through a horseshoe of coyote fur at this impromptu fête on top of the sea.

Engines roared against the wind's howl. We joined the convoy, twelve skidoos in all, sleds laden with tents, bedding rolls, supplies, red fuel cans, dead char and geese, lashed down under blue and orange tarps. Some of the qamutiiks carried long plywood boxes with women riding in them as if sitting up in their own coffins. The fog enveloped us, as white as the snow. The fixed relationships of ground and sky, vertical and horizontal, were suddenly effaced, leaving nothing but whiteness, as if we were in free fall, plummeting through cloud, feeling for ripcords. Each skidoo's headlamp probed the fog, light flashing off the silver reflector stripes on the backs of Snow Goose parkas. The headlamp beams linked one skidoo to the next, so that we travelled as a tube of light and heat through the fog, over the shorefast ice, and then inland, up and down hills, until we crested the last ridge of Mallik Island and saw, glimmering faintly, the lights of the town.

Next morning, the twelve-seater Beechcraft took off from Cape Dorset's airstrip and climbed over Hudson Strait. The ice had broken up: the water far below was deep blue, strewn with gleaming white scrims, plates and bergs. I was exhilarated. I had an end in sight. I wasn't patient. I wished the distance would collapse in a blink, a fingersnap. I flew from Iqaluit to Montreal and from Montreal to London, aware, minute by minute, of arrival closing in. I kept anticipating, leaping ahead, feeling my body lag behind. I couldn't concentrate on anything. I tried reading a book, but my mind wouldn't hold still, my attention wouldn't cleave to the lines.

The names of airport shops, the weight of coins, newspaper typefaces, voices, forms of address, the look of cars: I remembered these. I didn't take the train into the city. I caught a bus, the Flight, bound for the Midlands. It was June, midsummer, the trees in full leaf, the grass rich and luxuriant, and so much green, green everywhere, the whole country glutted with sap and pigment. The Flight cruised through the cutting in the Chiltern Hills, and on the far side, just as we came out into the Vale of Oxford, I saw a bird, a raptor, with white patches under its wings, and a deep-forked, rust-coloured tail. It held its wings steady, the black primary feathers at each tip spread like fingers, soaring on updrafts created as wind deflected off the north-facing slopes. I knew what it was. A red kite, *Milvus milvus* – I'd never seen one before, but I remembered my father telling me that a few red kites had been bred in captivity and released close to the cutting, that if you were lucky you might see one from the motorway, it had white patches under its wings, a rust-coloured belly and tail. I couldn't wait to tell him about the red kite. I wanted to tell everyone on the bus about the red kite.

At the bus station, when the taxi driver asked me where we were going, it took me by surprise, the pleasure of speaking the address, the shape of the words in my mouth. And then everything occurred in inevitable sequence: three roundabouts as we left the town, a red-bricked terraced row, a stand of Corsican pines, a school, and then signs, rooflines, the road's gradients and curves, the dairy buildings with their heavy sliding doors, the toll cottage, the almshouses, the fields in familiar patterns: Little Quarters, Morby's Close, the Shoulder of Mutton, the Great Ground. The colour of the ironstone. I looked over to the right, expecting to see cricket-bat willows along the Sor Brook, and there they were. Nothing had shifted.

The taxi turned right off the main road, slowed for an elderly woman out with her dog, crossed the Sor Brook by a stone bridge and pulled over at a passing place. I wanted to walk. I walked along the single-track road, carrying my two bags, a spring in my step. I could hear the brook rushing. Rooks were cawing. Sheep were grazing in Danvers Meadow. The spire appeared above the trees. The weathercock's tailplumes were glinting. I came to the yew by the churchyard gate. I saw the crowns of chestnuts, sycamores and limes, the white stone chimneys, the stone slate roof. Gravel crunched underfoot. Swallows swooped overhead. The rook caws grew louder as I walked up the drive to the house.

Author's Note

Thank you to the many people who showed me kindness on my journey to Foxe Land. Thank you to Deborah Rogers, Peter Straus, Laura Andreae, Mary Mount, Alicia Yerburgh, Dominic Oliver, Lydia Rainford, Rebecca Senior, Ann Godoff, Susanna Porter, Amanda Urban, Irène Andreae, Kate Wallis, Sonali Wijeyaratne, David Fitzherbert, Rebecca Stratford, Mark Espiner, Jane Kirkpatrick, Judy Bogdanor, Matt Ridley, Ulric Van den Bogaerde, Ingrid Wassenaar, Alex Monsey, Tom Bowring, Laurence Laluyaux and Stephen Edwards. For the sake of clarity, I have taken liberties with Captain Foxe's spelling when quoting from *The North-West Fox*. Quotations from *The Odyssey* come from the translation by Robert Fagles. *The Snow Geese* draws on the books and papers listed below.

Able, Kenneth P., and Verner P. Bingham. 1987. 'The Development of Orientation and Navigation Behavior in Birds.' *The Quarterly Review of Biology* 62:1–29.

Alerstam, Thomas. 1990. *Bird Migration*. Cambridge: Cambridge University Press.

Altringham, John D. 1996. *Bats: Biology and Behaviour*. Oxford: Oxford University Press.

Baker, R. Robin. 1982. *Migration: Paths through Time and Space*. London: Hodder & Stoughton.

——, ed. 1991. *Fantastic Journeys*. London: Merehurst.

Bartlett, Des and Jen. 1975. *The Flight of the Snow Geese*. New York: Stein & Day.

Batt, Bruce. 1998. *Snow Geese: Grandeur and Calamity on an Arctic Landscape*. Memphis: Ducks Unlimited, Inc.

Bellrose, F.C. 1981. *Ducks, Geese and Swans of North America*. Harrisburg: Stackpole Books.

Berthold, Peter, ed. 1991. *Orientation in Birds*. Basel: Birkhauser Verlag.

———. 1993. *Bird Migration*. Oxford: Oxford University Press.

Bickle, Ian. 1995. *Turmoil and Triumph: The Controversial Railway to Hudson Bay*. Calgary: Detselig Enterprises Ltd.

Bone, Neil. 1991. *The Aurora: Sun-Earth Interactions*. Chichester: John Wiley & Sons Ltd.

Brekke, Asgeir, and Alv Egeland. 1983. *The Northern Light*. Berlin: Springer Verlag.

Bromhall, Derek. 1980. *Devil Birds: The Life of the Swift*. London: Hutchinson & Co.

Bull, John, and John Farrand, Jr. 1998. *National Audubon Society Field Guide to North American Birds (Eastern Region)*. New York: Alfred A. Knopf.

Copland, James. 1858. *A Dictionary of Practical Medicine*. London: Longmans, Green, & Co.

Dickinson, Mary B., ed. 1999. *Field Guide to the Birds of North America* (Third Edition). Washington, D.C.: National Geographic Society.

Ehrlich, Paul R., David S. Dobkin and Darry Wheye. 1988. *The Birder's Handbook: A Field Guide to the Natural History of North American Birds*. New York: Simon & Schuster Inc.

Elphick, Jonathan, ed. 1995. *Collins Atlas of Bird Migration*. London: HarperCollins.

Emlen, S.T. 1967. 'Migratory Orientation in the Indigo Bunting, *Passerina cyanea*. I. The Evidence for Celestial Cues.' *Auk* 84:309–42.

———. 1967. 'Migratory Orientation in the Indigo Bunting, *Passerina cyanea*. II. Mechanisms of Celestial Orientation.' *Auk* 84:463–89.

———. 1975. 'Migration: Orientation and Navigation.' *Avian Biology* 5:129–219.

Fisher, Shirley. 1988. 'Leaving Home: Homesickness and the Psychological Effects of Change and Transition.' *Handbook of Life Stress, Cognition and Health*, ed. Fisher, S., and J. Reason. London: John Wiley & Sons Ltd.

Flicker, David J., and Paul Weiss. 1943. 'Nostalgia and its Military Implications.' *War Medicine* 4, 4:380–87.

Fodor, Nandor. 1950. 'Varieties of Nostalgia.' *Psychoanalytical Review* 37: 25–38.

Gallico, Paul. 1941. *The Snow Goose.* London: Michael Joseph.

Gauthreaux, S.A., Jr. 1982. 'The Ecology and Evolution of Avian Migration Systems.' *Avian Biology* 6:93–168.

Gill, Frank. 1990. *Ornithology.* New York: W.H. Freeman and Co.

Gwinner, E. 1977. 'Circannual Rhythms in Bird Migration.' *Annual Review of Ecology and Systematics* 8:381–405.

Hallendy, Norman. 1992. 'Inuksuit: Semalithic Figures Constructed by Inuit in the Canadian Arctic.' Paper prepared for the 25th Annual Meeting of the Canadian Archaeological Association, London, Ontario.

Henderson, Jim, and John MacNichol. 1997. *The Aurora.* Aboyne: Crooktree Images.

Hill, John E., and James D. Smith. 1984. *Bats: A Natural History.* London: British Museum.

Hofer, Johannes. 1688. 'Medical Dissertation on Nostalgia', trans. C.K. Anspach. *Bulletin of the Institute of the History of Medicine* 2:376–91, 1934.

Homer, trans. Robert Fagles. 1996. *The Odyssey*. New York: Viking Penguin.

Johnsgard, Paul A. 1991. *Crane Music*. Washington, D.C.: Smithsonian Institution Press.

Keeton, W.T. 1979. 'Avian Orientation and Navigation: A Brief Overview.' *British Birds* 72:451–70.

Kerlinger, Paul. 1995. *How Birds Migrate*. Mechanicsburg: Stackpole Books.

Kramer, Gustav. 1952. 'Experiments on Bird Orientation.' *Ibis* 94:265–85.

Kristjanson, Wilhelm. 1965. *The Icelandic People in Manitoba*. Winnipeg: Wallingford Press.

Lack, David. 1956. *Swifts in a Tower*. London: Methuen & Co.

Lopez, Barry. 1986. *Arctic Dreams*. New York: Charles Scribner's Sons.

Martin, Alexander R. 1954. 'Nostalgia.' *The American Journal of Psychoanalysis* 14:93–104.

Martin, Constance. 1995. *Search for the Blue Goose*. Calgary: Bayeux Arts Inc.

McCann, Willis H. 1941. 'Nostalgia: A Review of the Literature.' *Psychological Bulletin*, 38:165–82.

McEwan, Grant. 1975. *The Battle for the Bay*. Saskatoon: Western Producer Book Service.

McIlhenny, E.A. 1942. 'The Blue Goose in its Winter Home.' *Auk* 49:1278–1307.

Mead, Chris. 1983. *Bird Migration*. Feltham: Newnes Books.

Miller, Christy, ed. 1894. *The Voyages of Captain Luke Foxe of Hull, and Captain Thomas James of Bristol, in search of a northwest passage, in 1631–32, with narratives of the earlier northwest voyages of Frobisher, Davis, Weymouth, Hall, Knight,*

Hudson, Button, Gibbons, Bylot, Baffin, Hawkridge, and others.
London: Hakluyt Society.

Murray, W.H. 1981. *The Curling Companion.* Glasgow: Richard Drew Publishing.

Owen, M. 1980. *Wild Geese of the World: Their Life History and Ecology.* London: B.T.Batsford.

Perdeck, A.C. 1958. 'Two Types of Orientation in Migrating Starlings, *Sturnus vulgaris*, and Chaffinches, *Fringilla coelebs*, as Revealed by Displacement Experiments.' *Ardea* 46:1–37.

———. 1967. 'Orientation of Starlings after Displacement to Spain.' *Ardea* 55:194–202.

Peterson, Roger Tory, Guy Mountfort and P.A.D. Hollom. 1993. *Birds of Britain and Europe.* London: HarperCollins.

Petrie, William. 1963. *Keoeeit: The Story of the Aurora Borealis.* Oxford: Pergamon Press.

Robertson, Donna G., and R. Douglas Slack. 1995. 'Landscape Change and its Effects on the Wintering Range of a Lesser Snow Goose *Chen caerulescens caerulescens* Population: A Review.' *Biological Conservation* 71:179–85.

Rosen, George. 1975. 'Nostalgia: a "Forgotten" Psychological Disorder.' *Psychological Medicine* 5:340–54.

Rutstrum, Calvin. 1961. *The Wilderness Cabin.* New York: Macmillan.

Sauer, E.G.F. 1958. 'Celestial Navigation by Birds.' *Scientific American* 199:42–7.

Savage, Candace. 1994. *Aurora: The Mysterious Northern Lights.* Vancouver: Greystone Books.

Schmidly, David J. 1991. *The Bats of Texas.* College Station: Texas A&M University Press.

Soper, John Dewey. 1930. 'Discovery of the Breeding Grounds of the Blue Goose.' *The Canadian Field Naturalist* 44:1–11.

——. 1942. 'Life History of the Blue Goose.' *Proceedings of the Boston Society of Natural History* 42, 2:121–225.

Stresemann, Ernest. 1975. *Ornithology: From Aristotle to the Present.* Cambridge: Harvard University Press.

Thomson, A. Landsborough. 1936. *Bird Migration.* London: H. F. & G. Witherby Ltd.

Weidensaul, Scott. 1999. *Living on the Wind: Across the Hemisphere with Migratory Birds.* New York: North Point Press.

Werman, David S. 1977. 'Normal and Pathological Nostalgia.' *Journal of the American Psychoanalytical Association* 25, 2:387–98.

Wetmore, Alexander. 1926. *The Migrations of Birds.* Cambridge: Harvard University Press.

Wiltschko, W., and R. Wiltschko. 1972. 'Magnetic Compass of European Robins.' *Science* 176:62–4.

——. 1988. 'Magnetic Orientation in Birds.' *Current Ornithology* 5:67–121.

Van Tilburg, M.A.L., A.J.J.M Vingerhoets and G.L. Van Heck. 1996. 'Homesickness: A Review of the Literature.' *Psychological Medicine* 26:899–912.

Young, Steven B. 1989. *To The Arctic.* New York: John Wiley & Sons Ltd.